Hardhatting in a Geo-World

Authors
Betty Cordel
Barry Courtney
Helen Crossley
Susan Dixon
Geraldine Haracz
Loretta Hill
Ann Wiebe
Nancy Williams

Editors
Ann Wiebe
Betty Cordel

Illustrators
Reneé Mason
Margo Pocock
Brenda Richmond

Desktop Publisher
Tanya Adams

HARDHATTING IN A GEO-WORLD © 2004 AIMS Education Foundation

This book contains materials developed by the AIMS Education Foundation. **AIMS** (**A**ctivities **I**ntegrating **M**athematics and **S**cience) began in 1981 with a grant from the National Science Foundation. The non-profit AIMS Education Foundation publishes hands-on instructional materials (books and the quarterly magazine) that integrate curricular disciplines such as mathematics, science, language arts, and social studies. The Foundation sponsors a national program of professional development through which educators may gain both an understanding of the AIMS philosophy and expertise in teaching by integrated, hands-on methods.

Copyright © 1986, 1996, 2004 by the AIMS Education Foundation

All rights reserved. No part of this work may be reproduced or transmitted in any form or by any means—graphic, electronic, or mechanical, including photocopying, taping, or information storage/retrieval systems—without written permission of the publisher unless such copying is expressly permitted by federal copyright law. The following are exceptions to the foregoing statements:

- A person or school purchasing this AIMS publication is hereby granted permission to make up to 200 copies of any portion of it, provided these copies will be used for educational purposes and only at that school site.

- An educator presenting at a conference or providing a professional development workshop is granted permission to make one copy of any portion of this publication for each participant, provided the total does not exceed five activities per workshop or conference session.

Schools, school districts, and other non-profit educational agencies may purchase unlimited duplication rights for AIMS activities for use at one or more school sites. Visit the AIMS website, www.aimsedu.org, for more information or contact AIMS:

P.O. Box 8120, Fresno, CA 93747-8120 • 888.733.2467 • 559.255.6396 (fax) • aimsed@aimsedu.org

ISBN **978-1-932093-10-X**

Printed in the United States of America

Hardhatting in a Geo-World

Structures
Pillars of Strength 1
Working Out the Wiggles 7
Constructing With Straws..12
Straws Take a Stand 13
Sky High 18
Thanks for Your Support!...24
Bridge It 29

Measurement
Student-Made Measuring Tools 34

Length
- Rulers Line Up 37
- Links to Length 43
- Are You a Square? 48
- Bear Facts 54

Mass
- Cups 'n' Stuff..59
- Peddle the Metal 64

Volume
- Filling Stations 69
- Pleased as Punch 75

Time
- Minute Minders 81

Angle
- From Wedges to Wangles 86
- Waxed Wangles 94
- Wangle Round Up 101

Geometry
Shaping Up 111
Slice Me Twice..117
Möbius Bands 122
Geo-Panes..127
Edge to Edge 132
Net-Sense 139
Wreck-Tangles..146
Paper Pinchers 152
Circle Sighs 161
Playground Geometry..167
Once Around the Track 172

Meter Tape 178

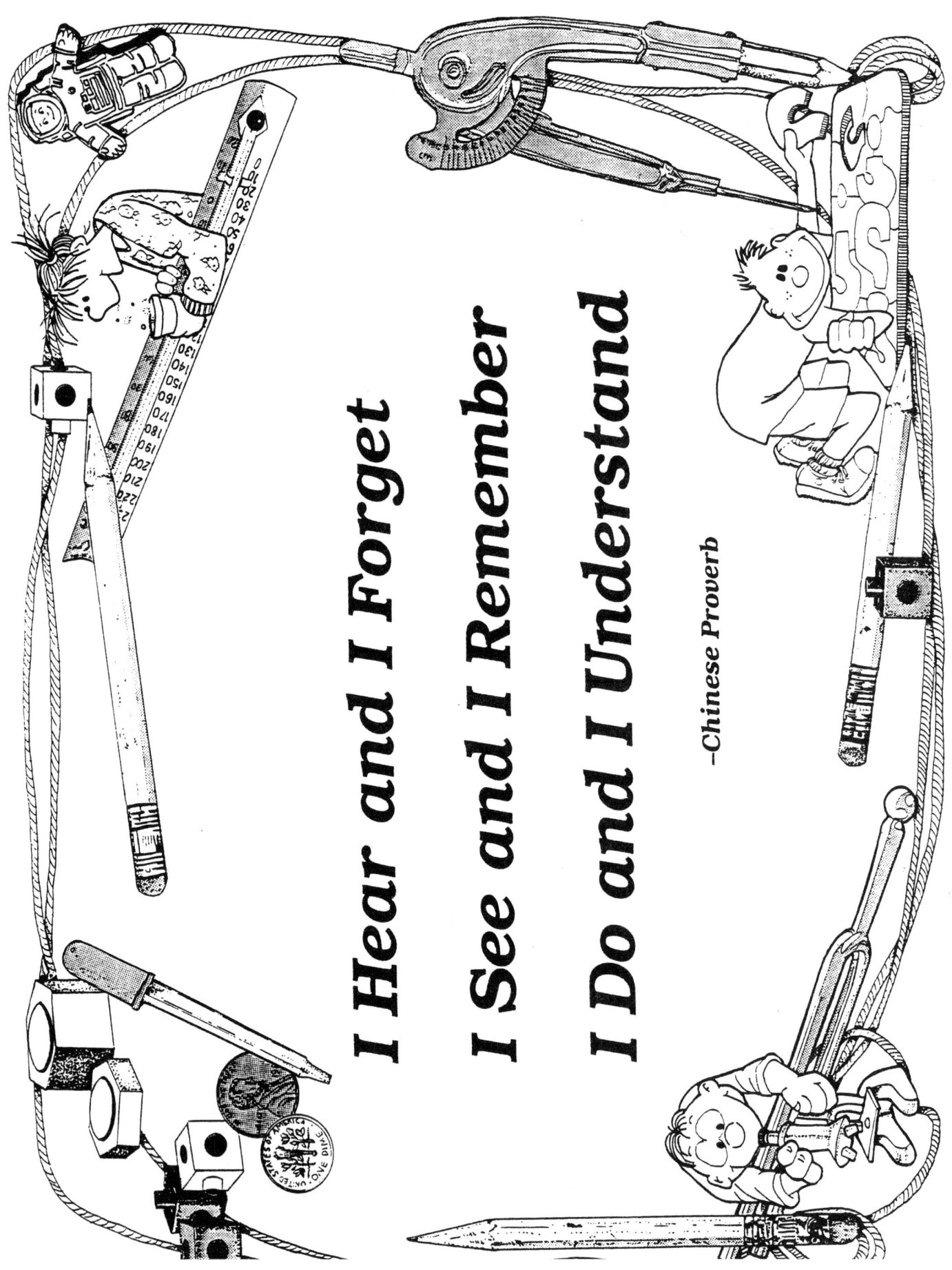

I Hear and I Forget

I See and I Remember

I Do and I Understand

–Chinese Proverb

National Reform Documents

NCTM Standards 2000*

Number and Operations
Compute fluently and make reasonable estimates
- Develop fluency in adding, subtracting, multiplying, and dividing whole numbers

Algebra
Understand patterns, relations, and functions
- Describe, extend, and make generalizations about geometric and numeric patterns

Geometry
Analyze characteristics and properties of two- and three-dimensional geometric shapes and develop mathematical arguments about geometric relationships
- Identify, compare, and analyze attributes of two- and three-dimensional shapes and develop vocabulary to describe the attributes
- Classify two- and three-dimensional shapes according to their properties and develop definitions of classes of shapes such as triangles and pyramids
- Investigate, describe, and reason about the results of subdividing, combining, and transforming shapes
- Explore congruence and similarity

Use visualization, spatial reasoning, and geometric modeling to solve problems
- Build and draw geometric objects
- Create and describe mental images of objects, patterns, and paths
- Identify and build a three-dimensional object from two-dimensional representations of that object
- Identify and build a two-dimensional representation of a three-dimensional object
- Use geometric models to solve problems in other areas of mathematics, such as number and measurement
- Recognize geometric ideas and relationships and apply them to other disciplines and to problems that arise in the classroom or in everyday life.

Measurement
Understand measurable attributes of objects and the units, systems, and processes of measurement
- Understand such attributes as length, area, weight, volume, and size of angle and select the appropriate type of unit for measuring each attribute
- Understand the need for measuring with standard units and become familiar with standard units in the customary and metric systems
- Understand that measurements are approximations and understand how differences in units affect precision
- Explore what happens to measurements of a two-dimensional shape such as its perimeter and area when the shape is changed in some way

Apply appropriate techniques, tools, and formulas to determine measurements
- Measure with multiple copies of units of the same size, such as paper clips laid end to end
- Select and use benchmarks to estimate measurements
- Select and apply appropriate standard units and tools to measure length, area, volume, weight, time, temperature, and the size of angles

Data
Formulate questions that can be addressed with data and collect, organize, and display relevant data to answer them
- Collect data using observations, surveys, and experiments
- Represent data using tables and graphs such as line plots, bar graphs, and line graphs

Problem Solving
- Build new mathematical knowledge through problem solving
- Solve problems that arise in mathematics and in other contexts

Representation
- Use representations to model and interpret physical, social, and mathematical phenomena

* Reprinted with permission from *Principles and Standards for School Mathematics, 2000* by the National Council of Teachers of Mathematics. All rights reserved.

Project 2061 Benchmarks*

The Nature of Science
- Results of scientific investigations are seldom exactly the same, but if the differences are large, it is important to try to figure out why. One reason for following directions carefully and for keeping records of one's work is to provide information on what might have caused the differences.

The Nature of Mathematics
- Mathematical ideas can be represented concretely, graphically, and symbolically.
- Mathematics is the study of many kinds of patterns, including numbers and shapes and operations on them. Sometimes patterns are studied because they help to explain how the world works or how to solve practical problems, sometimes because they are interesting in themselves.
- Numbers and shapes—and operations on them—help to describe and predict things about the world around us.

The Nature of Technology
- Measuring instruments can be used to gather accurate information for making scientific comparisons of objects and events and for designing and constructing things that will work properly.
- Even a good design may fail. Sometimes steps can be taken ahead of time to reduce the likelihood of failure, but it cannot be entirely eliminated.
- Scientific laws, engineering principles, properties of materials, and construction techniques must be taken into account in designing engineering solutions to problems. Other factors, such as cost, safety, appearance, environmental impact, and what will happen if the solution fails also must be considered.

The Mathematical World
- When people care about what is being counted or measured, it is important for them to say what the units are (three degrees Fahrenheit is different from three centimeters, three miles from three miles per hour).
- Measurements are always likely to give slightly different numbers, even if what is being measured stays the same.
- Tables and graphs can show how values of one quantity are related to values of another.
- Length can be thought of as unit lengths joined together, area as a collection of unit squares, and volume as a set of unit cubes.
- Graphical display of numbers may make it possible to spot patterns that are not otherwise obvious, such as comparative size and trends.
- Shapes such as circles, squares, and triangles can be used to describe many things that can be seen.
- Many objects can be described in terms of simple plane figures and solids. Shapes can be compared in terms of concepts such as parallel and perpendicular, congruence and similarity, and symmetry. Symmetry can be found by reflection, turns, or slides.
- Some shapes have special properties: Triangular shapes tend to make structures rigid, and round shapes give the least possible boundary for a given amount of interior area. Shapes can match exactly or have the same shape in different sizes.
- Spreading data out on a number line helps to see what the extremes are, where they pile up, and where the gaps are. A summary of data includes where the middle is and how much spread is around it.

Common Themes
- Some features of things may stay the same even when other features change. Some patterns look the same when they are shifted over, or turned, or reflected, or seen from different directions.

Habits of Mind
- Keep records of their investigations and observations and not change the records later.
- Offer reasons for their findings and consider reasons suggested by others.
- Add, subtract, multiply, and divide whole numbers mentally, on paper, and with a calculator.
- Judge whether measurements and computations of quantities such as length, area, volume, weight, or time are reasonable in a familiar context by comparing them to typical values.
- Assemble, describe, take apart and reassemble constructions using interlocking blocks, erector sets, and the like.
- Make something out of paper, cardboard, wood, plastic, metal, or existing objects that can actually be used to perform a task.
- Measure and mix dry and liquid materials (in the kitchen, garage, or laboratory) in prescribed amounts, exercising reasonable safety.
- Make sketches to aid in explaining procedures or ideas.
- Use numerical data in describing and comparing objects and events.

* American Association for the Advancement of Science. *Benchmarks for Science Literacy.* Oxford University Press. New York. 1993.

NRC Standards*

Science as Inquiry
- Plan and conduct a simple investigation.
- Employ simple equipment and tools to gather data and extend the senses.
- Think critically and logically to make the relationships between evidence and explanations.
- Communicate investigations and explanations.

Physical Science
- Objects have many observable properties, including size, weight, shape, color, temperature, and the ability to react with other substances. Those properties can be measured using tools, such as rulers, balances, and thermometers.

Science and Technology
- Identify a simple problem.
- Propose a solution.
- Implementing proposed solutions.
- Evaluate a product or design.
- Communicate a problem, design, and solution.
- Tools help scientists make better observations, measurements, and equipment for investigations. They help scientists see, measure, and do things that they could not otherwise see, measure, and do.

* National Research Council. *National Science Education Standards.* National Academy Press. Washington D.C. 1996.

Pillars of STRENGTH

Topic
Strength of paper tubes

Key Questions
1. How can we make a stronger paper tube?
2. Challenge: Build a paper tube, at least 3 cm tall, that will support a person.

Learning Goals
Students will:
• explore how height, diameter, and thickness affect the strength of a paper tube, and
• build a paper tube that will support a person.

Guiding Documents
Project 2061 Benchmarks
• *Measuring instruments can be used to gather accurate information for making scientific comparisons of objects and events and for designing and constructing things that will work properly.*
• *Make something out of paper, cardboard, wood, plastic, metal, or existing objects that can actually be used to perform a task.*
• *Make sketches to aid in explaining procedures or ideas.*

NRC Standards
• *Employ simple equipment and tools to gather data and extend the senses.*
• *Evaluate a product or design.*
• *Communicate a problem, design, and solution.*

NCTM Standard 2000*
• *Select and apply appropriate standard units and tools to measure length, area, volume, weight, time, temperature, and the size of angles*

Math
Estimation
Measurement
　length
Geometry and spatial sense

Science
Physical science
　force

Technology
Engineering
　structures

Integrated Processes
Observing
Collecting and recording data
Identifying and controlling variables
Comparing and contrasting
Generalizing
Applying

Materials
Used copy paper
Tape
Hardback books (see *Management*)
Metric rulers

Background Information
Tubes are hollow and light, yet resist bending or twisting. They can stand alone or be part of a larger framework. Tubes used for strength are found in nature (plant stems, bones, etc.) and made by people (bicycle frames, metal ladder rungs, buildings, freeways, bridges, etc.). They are used in both horizontal and vertical positions, singularly or bundled.

Variables of height, diameter, thickness, material used, position (horizontal or vertical), and whether single or bundled determine the degree of support possible. Thick walls have more structural strength than thin walls. A larger diameter provides greater support because the mass is distributed over a larger area. Theoretically, the height of the tubes would not matter if they could always be kept truly vertical; however, in the real world, a taller tube is more likely to be off-vertical. This weakens the structure and causes it to collapse.

Management
1. For this activity, the variables of position (vertical), kind of paper (copy paper), and number of tubes (single rather than multiple) are controlled. Students explore height, diameter, and thickness.
2. Collect used copy paper so that plenty is available.
3. It is important to use identical books for the strength test. Identify a particular textbook, dictionary, etc. to be used.
4. To conduct the strength test, construct a tube, place it vertically on a smooth, flat surface, and carefully center a book on top of it. Keep adding books, one at a time, until the tube is crushed. Count the number of books it supported, minus the one that caused it to fall. It is not a fair test if a book imbalance caused the tube to crush.

HARDHATTING IN A GEO-WORLD　　　　1　　　　© 2004 AIMS Education Foundation

5. Groups of two or three are suggested.
6. To control the variable of mass when doing *Step On It*, a single person (teacher or student) should be chosen for the final test of all the tubes. Center a book (or clipboard) over the tube and have a person stand on top of the book with one foot, using a chair or another person to maintain balance.
7. The activity is divided into two parts, exploration (*Pillars of Strength*) and application (*Step On It*). These can be done on two separate days or for a longer period on one day. Consider using *Step On It* for assessment.

The following approach is offered for those students ready for more independent exploration.

> *Open-ended:* Ask the *Key Question* and explain the materials available to the groups. Let them plan their own procedure for exploring and reporting on this question. Then challenge them to build a paper tube that will support a person.

Procedure
1. Hold up a piece of used copy paper and ask, "How can we make a tube out of this?" Have students demonstrate their ideas.
2. Ask the *Key Question*, "How can we make a stronger paper tube?"
3. Have students think of different ways to change the tube (variables) as they explore this question. Once the variables have been identified, distribute the first activity sheet.
4. Review together how to conduct the strength test. Demonstrate, if needed.
5. Guide the discussion about the first variable to be tested, height. Ask, "What other variables cannot change during this test?" [diameter, thickness] The class or each group will need to decide what those standards will be. It is best to initially work with a single thickness of paper.
6. Instruct students to record the estimate of the number of books their best tube (during the three sets of tests) will support.
7. Have students perform the tests and record their data. They should record their best test result next to the estimate. You may wish to pause between each test for a discussion about the previous test and controlling variables for the next test. For example, students may reason that they should use the most successful tube height as the controlled height during the diameter test.
8. Have the groups report and compare their results.
9. Based on the data they have gathered, challenge students to build a paper tube that will support a person.
10. Distribute the *Step On It* activity sheet and explain how it is to be completed. Turn students loose to be creative engineers.
11. Have the chosen person test each group's tube. Students should compare the results, both in appearance and in performance.

Connecting Learning
1. What different ways can we build tubes from a piece of paper?
2. You've just finished the _____ test. What was your strongest tube? Why do you think this is so?
3. Look at the information from all of your tests. What combination would you use to construct the strongest possible tube?
4. How does your group's information compare with others in the class?
5. How did you use the data from *Pillars of Strength* to design the tube that supports a person?
6. How do the different groups' tubes that support a person compare in the way they look? …in the way they perform their job?
7. How are tubes used in construction for support?
8. You are constructing a building with some tall pillars that need to support a lot of mass. Describe the pillars you would want to use.

Extensions
1. Have each group find the mass of the books supported by their strongest tube on the first activity sheet. This might involve some problem solving.
2. Try other kinds of paper for making strong tubes.
3. Investigate the strength of attaching several tubes together (bundling).
4. Have students devise a plan for exploring the strength of tubes used as horizontal beams.

* Reprinted with permission from *Principles and Standards for School Mathematics*, 2000 by the National Council of Teachers of Mathematics. All rights reserved.

Pillars of STRENGTH

Key Questions

1. How can we make a stronger paper tube?
2. Build a paper tube, at least 3 cm tall, that will support a person.

Learning Goals

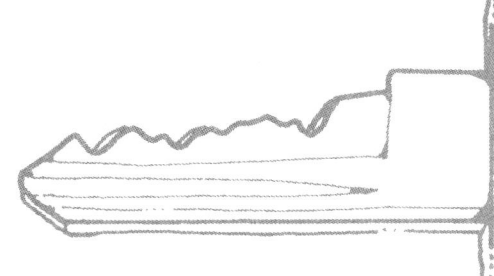

- explore how height, diameter, and thickness affect the strength of a paper tube, and
- build a paper tube that will support a person.

HARDHATTING IN A GEO-WORLD   3   © 2004 AIMS Education Foundation

Pillars of STRENGTH

Group
_____
_____
_____

How can we make a stronger paper tube?

The Strength Test
Place books on the tube, one at a time, until it collapses.

How many books will your **best** tube support?

Estimate: _____

Actual: _____

Test each variable and record the results below.

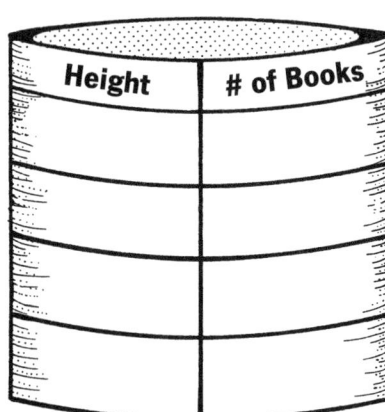

Height	# of Books

Keep the Same
Diameter: _____ cm
Thickness: 1 sheet

Diameter	# of Books

Keep the Same
Height: _____ cm
Thickness: 1 sheet

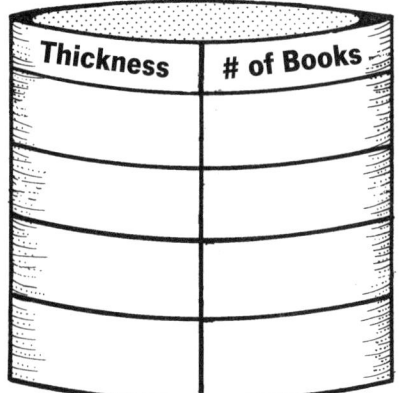

Thickness	# of Books

Keep the Same
Height: _____ cm
Diameter: _____ cm

HARDHATTING IN A GEO-WORLD © 2004 AIMS Education Foundation

Group _____

STEP ON IT

1. Build a paper tube, at least 3 cm tall, that will support a person.

2. Change and retest the tube until you are satisfied.

Construction Log
Draw and describe the tube each time it is changed.

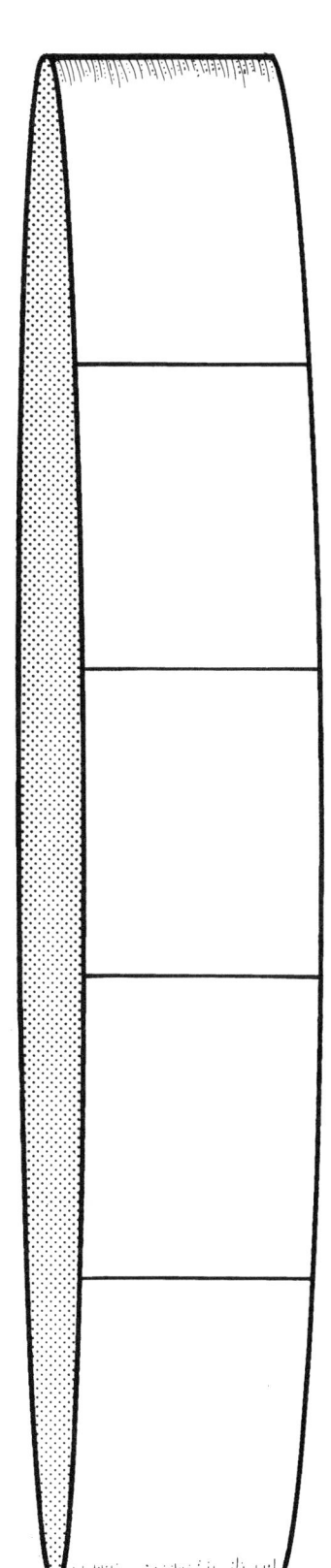

Compare your tube to others in the class.

HARDHATTING IN A GEO-WORLD 5 © 2004 AIMS Education Foundation

Pillars of STRENGTH

1. What different ways can we build tubes from a piece of paper?

2. You've just finished the _____ test. What was your strongest tube? Why do you think this is so?

3. Look at the information from all of your tests. What combination would you use to construct the strongest possible tube?

4. How does your group's information compare with others in the class?

5. How did you use the data from *Pillars of Strength* to design the tube that supports a person?

6. How do the different groups' tubes that support a person compare in the way they look? ...in the way they perform their job?

7. How are tubes used in construction for support?

8. You are constructing a building with some tall pillars that need to support a lot of mass. Describe the pillars you would want to use.

Topic
Stability of various polygons

Key Question
How can we make a structure stable?

Learning Goals
Students will:
- construct, test, and find ways to make polygons stable, and
- discover the shape that provides stability.

Guiding Documents
Project 2061 Benchmarks
- *Some shapes have special properties: Triangular shapes tend to make structures rigid, and round shapes give the least possible boundary for a given amount of interior area. Shapes can match exactly or have the same shape in different sizes.*
- *Assemble, describe, take apart and reassemble constructions using interlocking blocks, erector sets, and the like.*

NRC Standards
- *Identify a simple problem.*
- *Evaluate a product or design.*
- *Communicate a problem, design, and solution.*

NCTM Standards 2000*
- *Build and draw geometric objects*
- *Recognize geometric ideas and relationships and apply them to other disciplines and to problems that arise in the classroom or in everyday life.*
- *Use representations to model and interpret physical, social, and mathematical phenomena*

Math
Geometry
Problem solving

Science
Physical science
  force

Technology
Engineering
  structures

Integrated Processes
Observing
Collecting and recording data
Comparing and contrasting
Generalizing

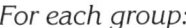

Materials
For the class:
  several hole punches
  scissors

For each group:
  6 tagboard strips, 2 cm x 10 cm
  3 tagboard strips, 2 cm x 20 cm
  6 paper fasteners

Background Information
The triangle is the only polygon that is rigid. All the sides push against each other to hold it in a stable position. The length of at least two sides would have to be changed in order for the triangle to move laterally. Triangles are commonly incorporated into the basic design of (or used to brace) buildings, towers, roofs, roller coasters, bridges, and many other kinds of structures made by people.

In this activity, students discover the stability of the triangle as they explore ways to make squares, pentagons, and hexagons rigid. Some possible solutions are shown below.

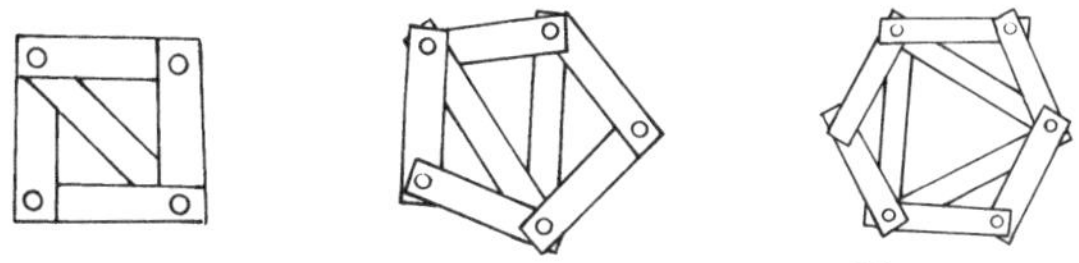

Square   Pentagon   Hexagon

Management
1. Construct one of the polygons beforehand to determine the size of paper fastener needed for the holes made by the hole punches. Small (1/8") hole punches work very well with No. 2, 1/2" paper fasteners. But use the hole punch size most easily available to you because the quantity of hole punches, in most cases, is more important than size in this activity.
2. If you have a limited number of hole punches, prepare the tagboard strips by punching holes in both ends of the 10-cm strips and in one end of the 20-cm strips.

HARDHATTING IN A GEO-WORLD   © 2004 AIMS Education Foundation

3. Have extra paper fasteners and 2 cm x 20 cm tagboard strips available as groups guess and test different bracing possibilities.
4. Groups of two work best.
5. Making a structure stable may involve several tries before success is achieved. Encourage students to persist until they reach their goal.

(The following is offered for those students ready for more independent investigation.)

> *Open-ended:* Illustrate the tagboard/paper fastener construction. Brainstorm shapes to try and discuss the meaning of *stable*. Introduce the *Key Question* and have students plan how they will record and report their discoveries. There should be evidence that they have studied the results and drawn conclusions. They might also generate questions to be answered.

Procedure
1. Join two 10-cm tagboard strips with a paper fastener and show this to the class. Explain that they will be building different shapes from these materials and finding which ones are stable. Tell them that *stable* means it will not move from side to side.
2. Give each group the tagboard strips, paper fasteners, and activity sheet. Have hole punches, extra strips, and extra paper fasteners available.
3. Instruct students to build one shape at a time with the 10-cm strips and draw a picture of it in the row labeled *Structure*.
4. Students should brace the shape, if needed, and complete the column in the table. To brace, they should attach a longer strip to the desired location, move it into position, mark where the second hole needs to be made, and punch a hole at this mark. Caution them not to trim the strip because they will use it again. (Bracing will probably be a trial-and-error experience. Encourage students to keep trying until they find a way that works.)
5. Direct students to take each structure apart and use the same materials for the next one.
6. When they have completed the table, give them time to study their drawings and answer the two questions. Let students determine a way to compare the number of sides and braces. (For example, they could make a table.)
7. Hold a concluding discussion.

Connecting Learning
1. What way(s) did you find to build the new shapes faster? [Don't take the whole shape apart. For example, just open up the triangle, add one more strip, and you have a square.]
2. When was bracing most difficult? [at the beginning (square) when I didn't know what worked, maybe at the end (hexagon) because it got more complicated]
3. How did the way you braced the square compare with other groups? Did more than one way work?
4. What is similar about all the square braces? [They divide the square into two regions, the endpoints are on adjoining strips.] Which brace do you think is strongest? [the one on the diagonal]
5. What ways did you try to brace the square that did not work? (Have students reflect on their guess-and-test process. Remind them that scientists experience many failures before they have a success.) *Repeat questions 3-5 for the pentagon and the hexagon.*
6. What do all the stable structures have in common? [They form triangles.]
7. How might an engineer use this information for a building, a tower, or a bridge? [They should use triangles when they design these structures.]
8. How do the number of sides and the number of braces compare? Is there a pattern? If so, what is it? (If students make a table, the pattern is easier to see: number of sides - 3 = number of braces.)

Extensions
1. Take the class on a walk around the neighborhood or the school grounds and look for examples of triangles being used to make stable structures. Have students observe triangles as they walk home from school.
2. Have students collect and examine pictures of bridges, towers, skyscrapers, playground equipment, etc. These may be made into a bulletin board collage.

Curriculum Correlation
Literature
Burns, Marilyn. *The Greedy Triangle*. Scholastic, Inc. New York. 1994.

Home Link
Have students search at home for items, such as toys, that use triangles for strength.

* Reprinted with permission from *Principles and Standards for School Mathematics,* 2000 by the National Council of Teachers of Mathematics. All rights reserved.

HARDHATTING IN A GEO-WORLD

Working Out the Wiggles

Key Question

How can we make a structure stable?

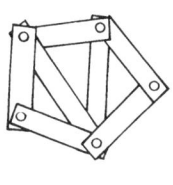 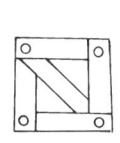 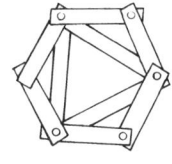

Learning Goals

Students will:

- construct, test, and find ways to make polygons stable, and
- discover the shape that provides stability.

Working Out the Wiggles

How can you make a structure stable?

Structure				
Shape	triangle	square	pentagon	hexagon
Is it stable?				
Show how you made it stable using the fewest braces.				

What do you notice about the stable structures?

How do the number of sides and the number of braces compare?

HARDHATTING IN A GEO-WORLD

Working Out the Wiggles

1. What way(s) did you find to build the new shapes faster?
2. When was bracing most difficult?
3. How did the way you braced the square compare with other groups? Did more than one way work?
4. What is similar about all the square braces? Which brace do you think is strongest?
5. What ways did you try to brace the ____ (square, pentagon, hexagon) that did not work?
6. What do all of the stable structures have in common?
7. How might an engineer use this information for a building, a tower, or a bridge?
8. How do the number of sides and the number of braces compare?

HARDHATTING IN A GEO-WORLD © 2004 AIMS Education Foundation

CONSTRUCTING WITH STRAWS

In order to build straw structures, the straws must be joined together. Several methods for joining are illustrated below. Choose the method that will best work for you, taking into consideration available materials, the manual dexterity of your students, the kind of project, and safety. A combination of methods may be needed for some projects.

INSERTION
Pinch the end of one straw and insert it into another straw. By making slits, straws can be joined in places other than at the ends.

PIPE CLEANERS
Bend short lengths of large pipe cleaners (chenille stems) and insert into the ends of straws. For in-between connections, cut a slit in the straw.

TRANSPARENT TAPE
Join straws by taping. A lot of tape will be needed.

PAPER CLIPS
Link two paper clips together and insert each clip into a straw. Paper clips can also be slid onto a straw for connections that are not on the ends of straws.

PINS
Attach straws by poking with straight pins. This is one of the fastest ways to build or change a structure, but it does raise a safety issue.

HARDHATTING IN A GEO-WORLD © 2004 AIMS Education Foundation

Straws Take a Stand

Topic
Stability of a cube

Key Question
How can you make a cube stable?

Learning Goal
Students will build a stable cube with straws.

Guiding Documents
Project 2061 Benchmarks
- Some shapes have special properties: Triangular shapes tend to make structures rigid, and round shapes give the least possible boundary for a given amount of interior area. Shapes can match exactly or have the same shape in different sizes.
- Assemble, describe, take apart and reassemble constructions using interlocking blocks, erector sets, and the like.

NRC Standards
- Identify a simple problem.
- Propose a solution.
- Implementing proposed solutions.

NCTM Standards 2000*
- Build and draw geometric objects
- Recognize geometric ideas and relationships and apply them to other disciplines and to problems that arise in the classroom or in everyday life
- Use representations to model and interpret physical, social, and mathematical phenomena

Math
Geometry
Problem solving

Science
Physical science
    force

Technology
Engineering
    structures

Integrated Processes
Observing
Comparing and contrasting
Generalizing

Materials
For each group:
    12 straws
    36 paper clips
    scissors

Background Information
The triangle is the basis for stable two- and three-dimensional shapes. It is the only polygon that is rigid. Its position cannot be altered unless the length of a side is changed. The triangle is commonly used in the design and bracing of buildings, towers, roofs, bridges, roller coasters, and many other kinds of structures.

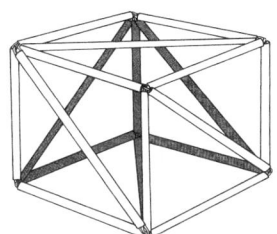

When the cube is constructed in this activity, students will experience its lack of stability. Through trial-and-error experimentation or the application of knowledge previously gained in *Working Out the Wiggles*, students will discover that triangular braces make the cube rigid.

The building of the tetrahedron intentionally follows the building of the cube. After having to add several braces to make the cube stable, students find the tetrahedron needs no such bracing. What's going on here? There may be a pause, then the moment of realization. The tetrahedron is made of triangles! Ahhh...the joy of discovery.

Management
1. Use groups of two or three. Responsibilities might be divided as follows: holding the structure, handling the paper clips, and inserting the straws.
2. For each group, cut six straws in half and leave the other six full-length.
3. Joining straws together can be challenging. One of the easier ways is to join two paper clips and insert each clip into a straw (*Diagram A*). If another paper clip is needed at that vertex, just attach it to the others. The paper clips should fit snugly inside the straws. Since the clips will not be bent open, they can be used for their normal purpose after this activity. Students who wish to try braces other than the diagonal can use the construction technique illustrated in *Diagram B*.

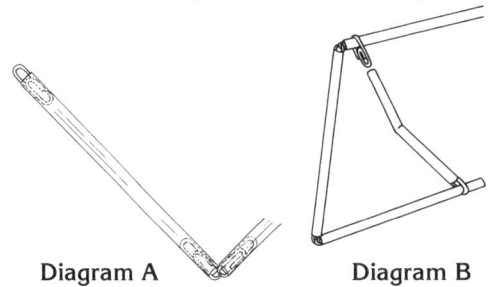

Diagram A   Diagram B

HARDHATTING IN A GEO-WORLD   © 2004 AIMS Education Foundation

Procedure

1. Issue the challenge: Build a stable cube with straws. Explain that stable means that the structure will not move from side to side; it is stiff. Demonstrate the method for joining straws together with paper clips.
2. Distribute the activity sheet, 12 half-length straws, and a starter pile of paper clips to each group. (Don't give students all 36 paper clips as this might provide hints in meeting the challenge. They can get more as they discover the need for them.)
3. Instruct students to build the cube.
4. Ask, "How stable is your cube?" [It is not stable at all.] "Does this surprise you?"
5. Have students continue with the challenge of making the cube stable. Make full-length straws and more paper clips available as they experiment with bracing. Students should cut these additional straws to the length needed.
6. When groups have met the challenge, show them how to draw a cube on the activity sheet.

Connect four points to make a square. From each corner, move up and over a chosen amount (such as 1 up, 1 over) and draw a second square.

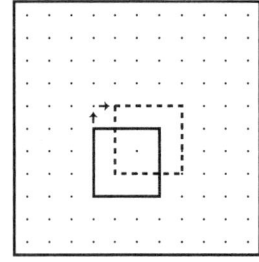

Connect the two squares by drawing lines between like corners.

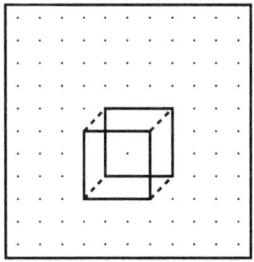

Add brace lines. (Not all brace lines are shown.)

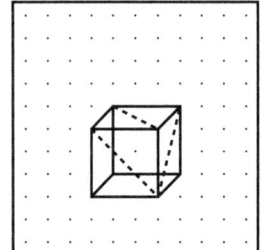

7. Have students record the 2-D shapes seen in their stable cubes.
8. Instruct students to take their cubes apart and build tetrahedrons (illustrated on activity sheet) with the same materials.
9. Hold a concluding discussion and have students generalize by describing their discovery.

Connecting Learning

1. What helpful construction techniques did you discover?
2. What surprised you? (possibly that the cube was so wobbly without braces, that the tetrahedron didn't need to be braced, etc.)
3. How do the cube and the tetrahedron compare? [One has six sides, the other four sides. The cube was not stable without braces, the tetrahedron was.]
4. What makes a structure stable? [triangles, whether braced or as a part of the original shape]
5. How many triangles were in the cube? [12] How does this relate to the number of sides (faces)? [It is double the number of sides because each brace divides a face into two sections.]
6. How would knowing that triangles make strong structures be helpful to people? [When planning a building, bridge, or tower, you want it to be safe for people to use and to do the job for which it was designed.]
7. Can you think of anything at your home that would benefit from being cross-braced? [bookshelves, etc.]
8. You are in charge of getting the materials ready for this activity. How many straws and paper clips do you need for our class? (Students will need to determine the number of groups, then count the materials they used to build and brace the cube. They should remember that some of the straws are cut in half so two half straws started out as one whole straw.)

Extension

Have the class search for structures around them that use triangles for strength: bicycles, playground equipment, etc.

* Reprinted with permission from *Principles and Standards for School Mathematics,* 2000 by the National Council of Teachers of Mathematics. All rights reserved.

HARDHATTING IN A GEO-WORLD

Straws Take a Stand

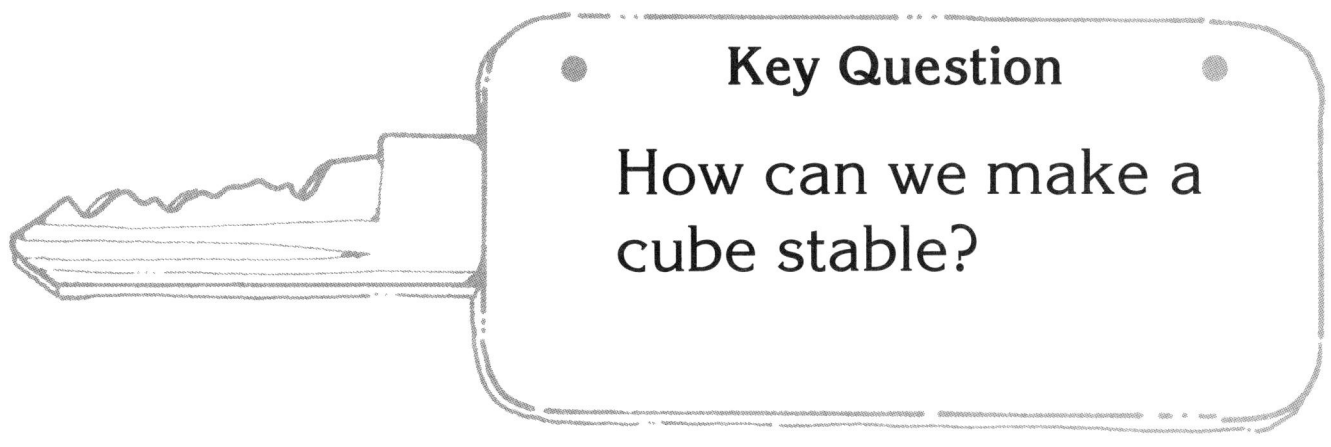

Key Question

How can we make a cube stable?

Learning Goal

Students will:

- build a stable cube with straws.

HARDHATTING IN A GEO-WORLD 15 © 2004 AIMS Education Foundation

Straws Take a Stand

1. What helpful construction techniques did you discover?

2. What surprised you?

3. How do the cube and the tetrahedron compare?

4. What makes a structure stable?

5. How many triangles were in the cube? How does this relate to the number of sides (faces)?

6. How would knowing that triangles make strong structures be helpful to people?

7. Can you think of anything at your home that would benefit from being cross-braced?

8. You are in charge of getting the materials ready for this activity. How many straws and paper clips do you need for our class?

Topic
Tall structures

Challenge
Build the tallest structure you can with 30 drinking straws.

Learning Goal
Students will use creativity, teamwork, and problem solving to build a tall, stable structure.

Guiding Documents
Project 2061 Benchmark
- Scientific laws, engineering principles, properties of materials, and construction techniques must be taken into account in designing engineering solutions to problems. Other factors, such as cost, safety, appearance, environmental impact, and what will happen if the solution fails also must be considered.

NRC Standards
- Identify a simple problem.
- Propose a solution.
- Implementing proposed solutions.
- Evaluate a product or design.

NCTM Standards 2000*
- Select and apply appropriate standard units and tools to measure length, area, volume, weight, time, temperature, and the size of angles
- Represent data using tables and graphs such as line plots, bar graphs, and line graphs

Math
Geometry and spatial sense
Measurement
  linear
Graphing
Median average

Science
Physical science
  force

Technology
Engineering
  structures

Integrated Processes
Observing
Predicting
Collecting and recording data
Comparing and contrasting
Applying

Materials
For each group:
  30 plastic straws
  material to join straws (see *Management*)
  clay to anchor structures
  meter sticks

Background Information
This is one of several activities in which students of all ages can apply the knowledge gained from *Working Out the Wiggles* and *Straws Take a Stand*. Triangular braces add strength. They are needed to make a strong base as well as to support the straw structure as it rises higher and higher.

If the data are ordered, in this case from shortest height to tallest height, the height in the middle is the median average. This kind of average is more easily understood than the mean average, particularly if the data are presented visually in a graph.

Management
1. Allow 1 to 1 1/2 hours. You may wish to set a time limit of 45-60 minutes for actual construction.
2. Groups of two or three should work together on a structure.
3. Construction should take place on the floor. Use clay to anchor the straw base to the floor. If working on carpet, clay is not needed.

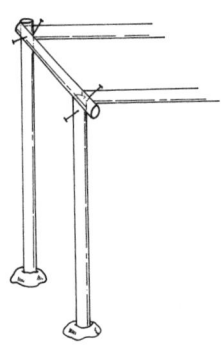

HARDHATTING IN A GEO-WORLD

4. Choose one or more methods for connecting straws from the options suggested in *Constructing With Straws* (see *Table of Contents*). Assemble the appropriate materials.

Procedure
1. Issue the *Challenge* and distribute the first activity page to each group (or each student). Have them predict the height of the tallest structure they think will be built.
2. Instruct a member of each group to collect the straws, clay, and other items needed for construction.
3. Show students how to anchor the clay and join straws together.
4. Have students build their structures.
5. Direct students to record their structure's estimated height, then measure its actual height.
6. Tell students to make a drawing of their structure and compare it to the tallest one built.
7. Guide the reporting of height data for all of the structures so students can record on the second activity page.
8. Have students make a bar graph, ordering the data from shortest to tallest. They should mark the middle height to find the median average.
9. Discuss the results.

Connecting Learning
1. How did you change (modify) your structure?
2. How did you use ideas you learned earlier?
3. What advice would you give someone who was going to build a similar structure? (You need a firmly-braced base or foundation. Use triangles whenever possible to strengthen the structure, etc.)
4. What surprised you about the different structures that were built?
5. Do you think a higher straw structure (than the tallest in class) could be built with only 30 straws? How would you plan to do it?
6. If you were to build another straw structure, what would you want the challenge to be?

* Reprinted with permission from *Principles and Standards for School Mathematics, 2000* by the National Council of Teachers of Mathematics. All rights reserved.

Challenge

Build the tallest structure you can with 30 drinking straws.

Learning Goal

Students will:

- use creativity, teamwork, and problem solving to build a tall, stable structure.

Sky High

Challenge: Build the tallest structure you can with 30 straws.

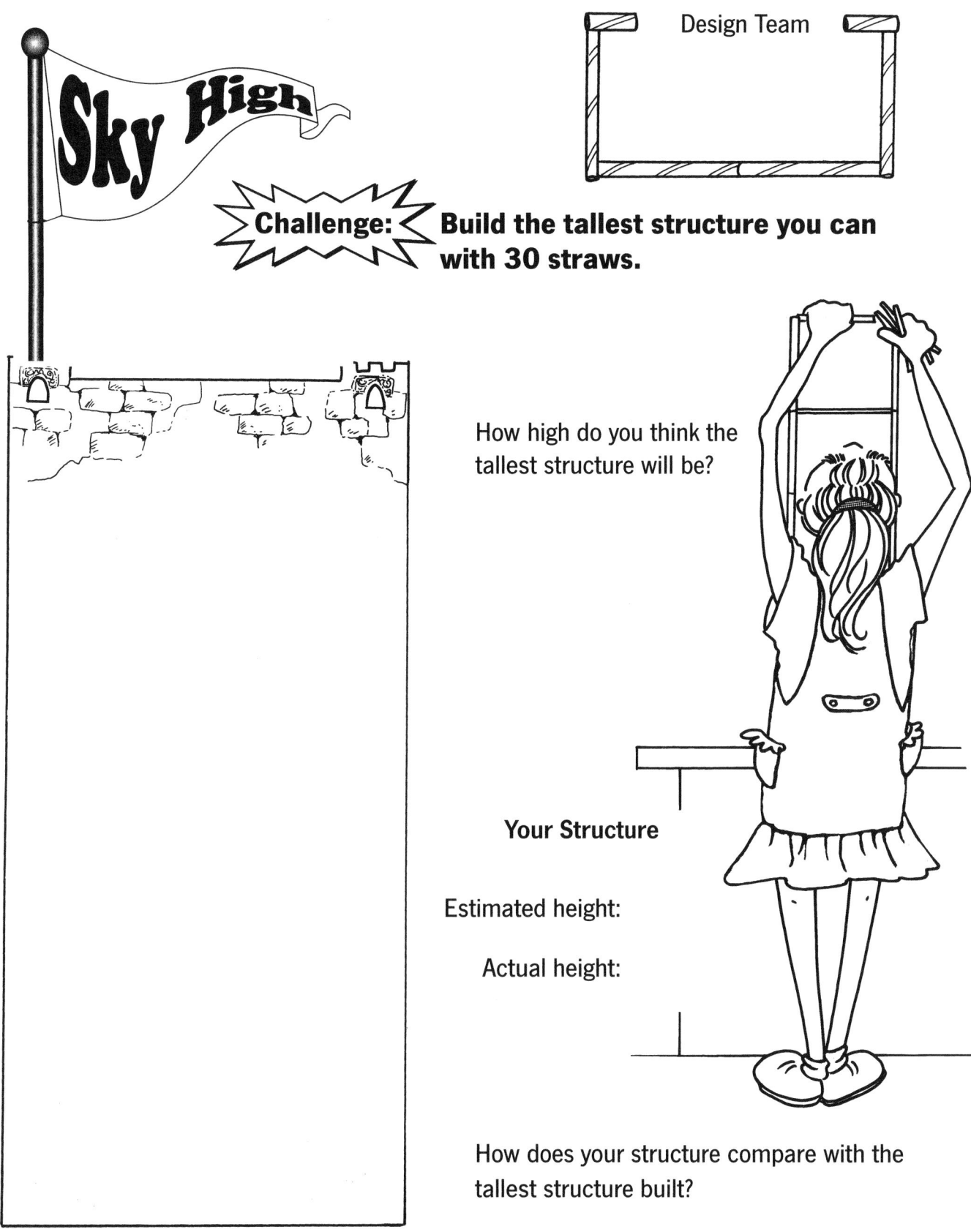

Design Team

How high do you think the tallest structure will be?

Your Structure

Estimated height:

Actual height:

How does your structure compare with the tallest structure built?

Draw your structure.

HARDHATTING IN A GEO-WORLD · 21 · © 2004 AIMS Education Foundation

Design Team

Record the heights of all the structures.

Graph the heights in order from shortest to tallest. Mark the middle height, the median average.

HARDHATTING IN A GEO-WORLD

Sky High

1. How did you change (modify) your structure?

2. How did you use ideas you learned earlier?

3. What advice would you give someone who was going to build a similar structure?

4. What surprised you about the different structures that were built?

5. Do you think a higher straw structure (than the tallest in class) could be built with only 30 straws? How would you plan to do it?

6. If you were to build another straw structure, what would you want the challenge to be?

Thanks For Your Support!

Topic
Supporting structures

Challenge
Build a structure that will support 400 grams at least 25 cm above the base.

Learning Goal
Students will use creativity, teamwork, and problem solving skills to build a straw structure that can hold 400 grams of mass.

Guiding Documents
Project 2061 Benchmarks
- *Scientific laws, engineering principles, properties of materials, and construction techniques must be taken into account in designing engineering solutions to problems. Other factors, such as cost, safety, appearance, environmental impact, and what will happen if the solution fails also must be considered.*
- *Even a good design may fail. Sometimes steps can be taken ahead of time to reduce the likelihood of failure, but it cannot be entirely eliminated.*

NRC Standards
- *Identify a simple problem.*
- *Propose a solution.*
- *Implementing proposed solutions.*
- *Evaluate a product or design.*
- *Communicate a problem, design, and solution.*

NCTM Standards 2000*
- *Solve problems that arise in mathematics and in other contexts*
- *Select and apply appropriate standard units and tools to measure length, area, volume, weight, time, temperature, and the size of angles*

Math
Geometry and spatial sense
Measurement
    linear
    mass

Science
Physical science
    force

Technology
Engineering
    structures

Integrated Processes
Observing
Collecting and recording data
Comparing and contrasting
Applying

Materials
For each group:
    30 plastic straws, minimum
    material to join straws (see *Management 4*)
    metric ruler
    empty 16 oz. sour cream or cottage cheese container
    material with a mass of 400 grams (see *Management 5*)

For the class:
    balance
    gram masses

Background Information
This is one of a series of activities that uses plastic drinking straws as the construction medium. The challenge presented requires creativity, cooperation, and problem solving skills. Students have an opportunity to apply what they have learned in *Working Out the Wiggles* and *Straws Take a Stand*, that triangular braces add strength to a structure. They will also evaluate their structures for cost, safety, and appearance.

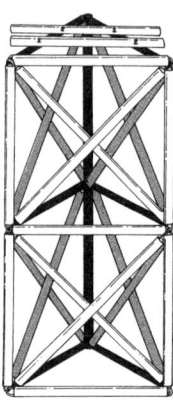
One solution

HARDHATTING IN A GEO-WORLD 24 © 2004 AIMS Education Foundation

Management
1. *Working Out the Wiggles* and *Straws Take a Stand* are strongly recommended preliminary activities.
2. Divide the class into groups of two or three.
3. Each group should start with 30 straws but have plenty of extras on hand in case they need more. If you wish to set a maximum, it should be at least 40 straws.
4. Choose one or more methods for connecting straws from the options suggested in *Constructing With Straws* (see *Table of Contents*). Assemble the appropriate materials.
5. Collect rocks or other materials that will fit into a sour cream container. Beforehand, measure or have each group measure 400 grams of rocks using the balance and gram masses. The rocks may be a bit over, but not under 400 grams. The rocks will gradually be added to the sour cream container to test the structure.

Procedure
1. Present students with the *Challenge*. Distribute the activity sheet, straws and other materials.
2. Review with the class how to test their structure. Carefully place the container on top of the structure and gently add a few rocks at a time. Solve any weakness problems before adding more mass.
3. Have students build and test their structures.
4. Direct each student or student group to measure the height and amount of mass supported, then draw the structure they built.
5. Ask students to look at the structures of other groups and compare with their own.
6. Discuss the results.

Connecting Learning
1. What changes did you make in your structure as you were building?
2. What do you like about your structure?
3. How did you use ideas you learned earlier?
4. How well did your group work together? Give an example.
5. Which structure has the most pleasing appearance?
6. Which structure appears to be the safest?
7. If each straw costs 10¢, how much did your structure cost to build? (You might assign prices to the other building materials used and have them figure total costs.)
8. If you were on a budget, for which structure would you be willing to pay? Why?

Curriculum Correlation
Literature
Wilson, Forrest. *What It Feels Like To Be A Building.* The Preservation Press, National Trust for Historic Preservation, 1785 Massachusetts Avenue. N.W., Washington, D.C. 20036. 1988. (This appealing book uses a minimum of words to show the push and pull of forces on a building.)

* Reprinted with permission from *Principles and Standards for School Mathematics, 2000* by the National Council of Teachers of Mathematics. All rights reserved.

HARDHATTING IN A GEO-WORLD

Challenge

Build a structure that will support 400 grams at least 25 cm above the base.

Learning Goal

Students will:

- use creativity, teamwork, and problem solving skills to build a straw structure that can hold 400 grams of mass.

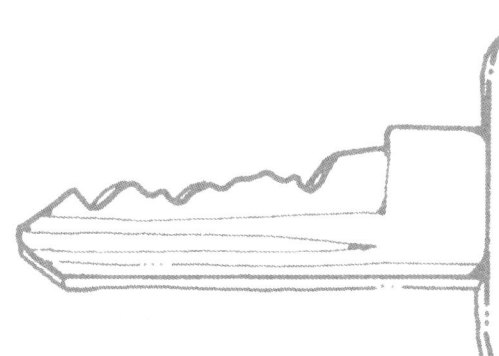

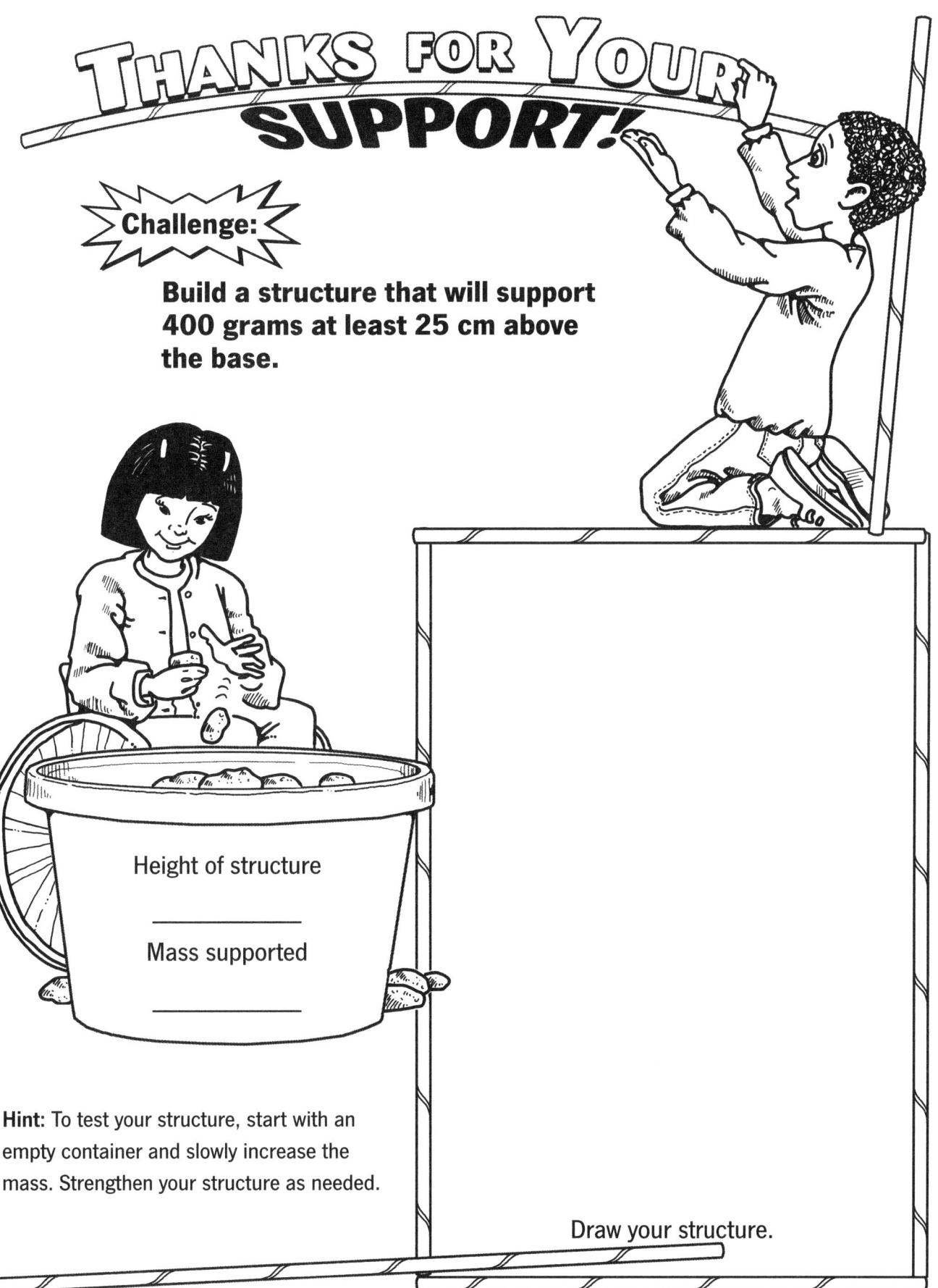

Thanks for Your Support!

1. What changes did you make in your structure as you were building?

2. What do you like about your structure?

3. How did you use ideas you learned earlier?

4. How well did your group work together? Give an example.

5. Which structure has the most pleasing appearance?

6. Which structure appears to be the safest?

7. If each straw costs 10¢, how much did your structure cost to build?

8. If you were on a budget, for which structure would you be willing to pay? Why?

BRIDGE IT

Topic
Bridges

Challenge
Build a straw bridge that spans a 30-cm gap and supports ____ grams on its road bed.

Learning Goal
Students will work as a team to meet the challenge of building a straw bridge to certain specifications.

Guiding Documents
Project 2061 Benchmarks
- Scientific laws, engineering principles, properties of materials, and construction techniques must be taken into account in designing engineering solutions to problems. Other factors, such as cost, safety, appearance, environmental impact, and what will happen if the solution fails also must be considered.
- Even a good design may fail. Sometimes steps can be taken ahead of time to reduce the likelihood of failure, but it cannot be entirely eliminated.

NRC Standards
- Identify a simple problem.
- Propose a solution.
- Implementing proposed solutions.
- Evaluate a product or design.
- Communicate a problem, design, and solution.

NCTM Standards 2000*
- Solve problems that arise in mathematics and in other contexts
- Select and apply appropriate standard units and tools to measure length, area, volume, weight, time, temperature, and the size of angles

Math
Geometry and spatial sense
Measurement
 linear
 mass

Science
Physical science
 force

Technology
Engineering
 structures

Integrated Processes
Observing
Collecting and recording data
Comparing and contrasting
Applying

Materials
For each group:
 30 plastic drinking straws, minimum
 material to join straws (see *Management 2*)
 beverage, fruit, vegetable, or soup can
  (see *Management 4*)
 meter stick, meter tape, or metric rulers
 scissors

For the class:
 balance
 gram masses

Background Information
Of the many kinds of bridges in the world, the truss bridge seems a logical choice when constructing with straws. Truss bridges use triangles for structural support. The triangular braces add strength to a structure. Experience with this concept, gained in *Working Out the Wiggles* and *Straws Take a Stand*, can and should be applied to this bridge-building task.

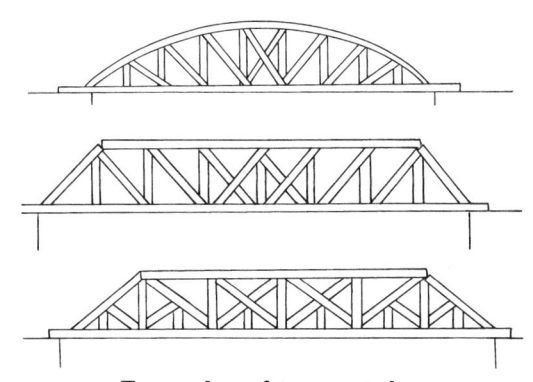

Examples of truss styles

The practical purpose for a bridge is to allow people or various forms of transportation to cross a gap. The road bed is the part of the bridge on which cars, trains, etc. actually travel. The bridge's ability to support mass on the road bed must be proven and is the place it is tested in this activity.

Aside from the engineering aspects, much of the value of this activity lies in challenging students' problem-solving skills and creativity along with developing cooperation and perseverance.

HARDHATTING IN A GEO-WORLD

One solution

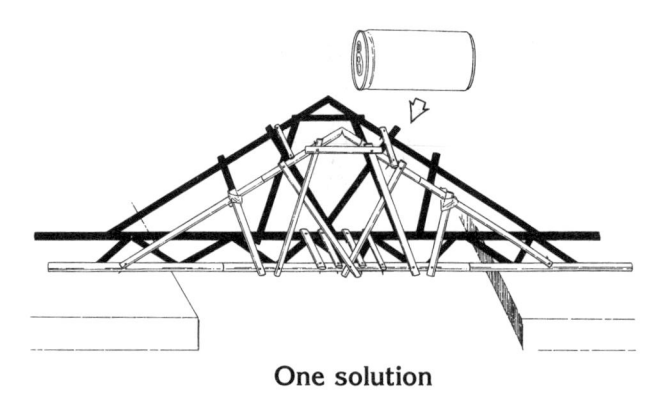

Management
1. *Working Out the Wiggles* and *Straws Take a Stand* are highly recommended preliminary activities.
2. A combination of methods will be needed to connect the straws in this activity. Choose from the options suggested in *Constructing With Straws* (see Table of Contents) and assemble the appropriate materials.
3. Students should work in groups of two or three.
4. Measure the mass of several canned products, choosing one between 300 and 500 grams. (The milliliters stated on cans cannot be equated with grams of mass.) Its mass, minus 10 grams or so, will determine the number to write in the Challenge. For a 380-gram soda can, you might write 370 grams. The goal should be slightly less than the actual mass used for testing since similar cans vary slightly.

   To test a minimum, you can't go under the amount and it is hard to be equal to the amount, so it is better to be slightly over the amount. Food processors add a little extra amount of food to their packages so the total mass will not fall below the minimum stated on the label.

   The opposite can also be true. If a bridge sign says, "2-ton limit", you would expect it can hold slightly more than that. The engineers allow for a little extra weight.
5. Encourage students to sketch a preliminary design on the back of the paper before beginning.
6. To help students keep track of the number of straws they use, you may wish to set up a check-out system. They could record the number they receive at the beginning and each time they get more. When the project is completed, they can subtract the straws they didn't use from the total number of straws checked out.

Procedure
1. Present the *Challenge*. Distribute the activity page and building materials. Explain various methods of joining straws, if needed.
2. Instruct each group to build their bridge across a 30-cm span between two flat desks or tables. The bridge cannot be attached to the desks.
3. Describe how to test the strength of the bridge without making it collapse. Carefully lower the side of the can to the center of the road bed. Without letting go, slowly let the bridge take more of the can's mass. If the bridge starts to give way, lift the can up, strengthen the bridge, and test again until the can rests on it without being held.
4. Let the groups begin construction.
5. Have each group complete the construction report and sketch their bridge.
6. Guide students on a room tour of the completed bridges and hold a concluding discussion.

Connecting Learning
1. How did you decide where to start? Did you draw a design first? Did you talk it over? Did you just start and see what would happen?
2. What problems did you have in meeting the challenge? What did you do about them?
3. What do you like about your bridge?
4. How many triangles does your bridge have?
5. How does your bridge compare with others? (Awards might be given for the most unusual bridge, the bridge built with the least amount of straws, the most pleasing bridge, the safest bridge, etc.)

Curriculum Correlation
Resources

Use these books to further explore bridge designs. Which bridges could be made with straws? What other materials could be used for model bridges?

Carter, Polly. *The Bridge Book*. Simon & Schuster. New York. 1992. (A historical look at the development of bridges, with emphasis on invention and the problem-solving process, delivered through cartoons and text.)

Robbins, Ken. *Bridges*. Dial Books. New York. 1991. (Beautifully written text with uniquely-colored drawings based on actual photographs of bridges in northeastern United States.)

Wilson, Forrest. *Bridges Go From Here To There*. The Preservation Press, National Trust for Historic Preservation, 1785 Massachusetts Avenue, N.W., Washington, D.C. 20036. 1993. (Clever black-and-white illustrations show the push and pull of forces on various kinds of bridges. Minimal words.)

* Reprinted with permission from *Principles and Standards for School Mathematics, 2000* by the National Council of Teachers of Mathematics. All rights reserved.

BRIDGE IT

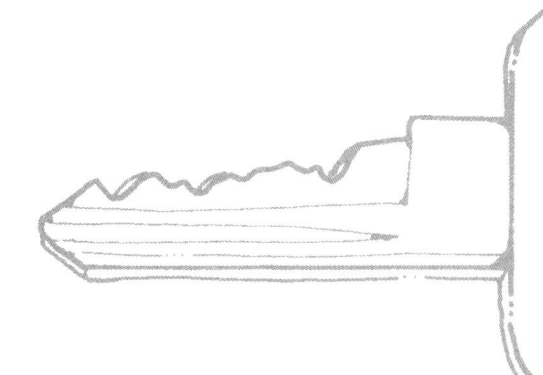

Challenge

Build a straw bridge that spans a 30 cm gap and supports ____ grams on its road bed.

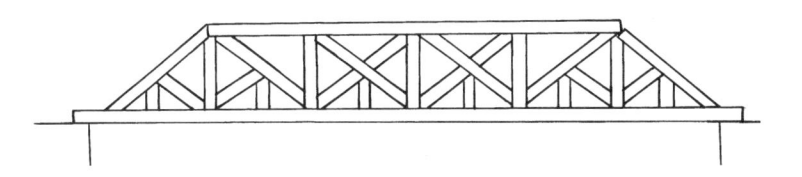

Learning Goal

Students will:

- work as a team to meet the challenge of building a straw bridge to certain specifications.

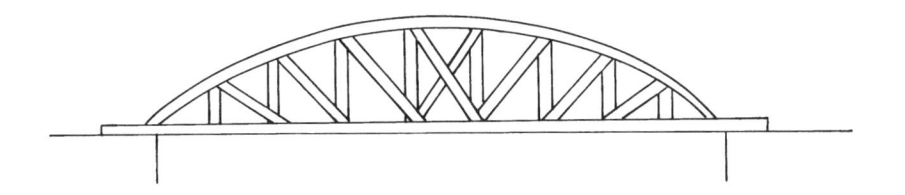

HARDHATTING IN A GEO-WORLD

BRIDGE IT

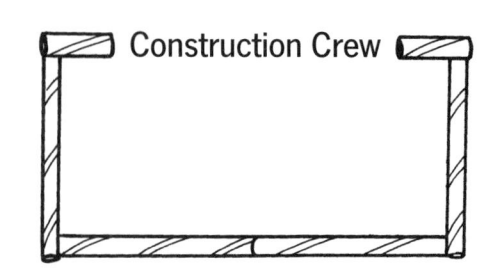

Construction Crew

Challenge:

Build a straw bridge that spans a 30-cm gap and supports _____ grams on its road bed.

Construction Report

Methods used to join straws:

Number of straws:

Length of bridge:

Width of bridge:

Mass supported:

Sketch your bridge.

HARDHATTING IN A GEO-WORLD   32   © 2004 AIMS Education Foundation

BRIDGE IT

1. How did you decide where to start? Did you draw a design first? Did you talk it over? Did you just start and see what would happen?

2. What problems did you have in meeting the challenge? What did you do about them?

3. What do you like about your bridge?

4. How many triangles does your bridge have?

5. How does your bridge compare with others?

Student-Made Measuring Tools

It is well worth the time for students to make their own measuring tools. Handmade tools tend to be well cared for because students feel ownership of them. Instructions for making these tools follow.

Linear Measure

Paper meter tapes

Goal: 2 per student, each a different color

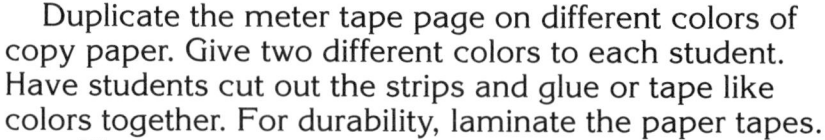

Duplicate the meter tape page on different colors of copy paper. Give two different colors to each student. Have students cut out the strips and glue or tape like colors together. For durability, laminate the paper tapes.

The meter tapes can be used singly for activities measuring short distances. To measure long distances, have students combine alternately-colored meter tapes to create a visually easy-to-read measure even at a distance.

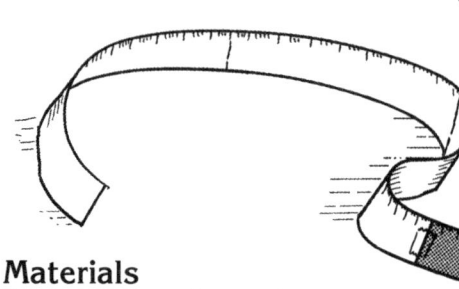

Materials
2 or more colors of copy paper
Scissors
Glue or tape

String meter measure

Goal: 1 per group

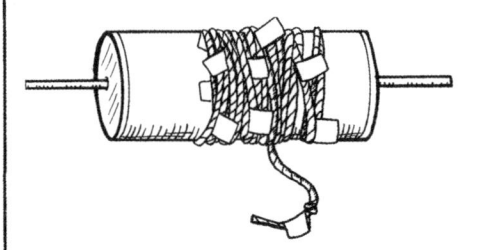

Make a knot a few centimeters from one end of the string; this represents zero. Use a meter tape to mark the one-meter point on the string with the permanent marker. (Both the tape and the string must be taut.) Continue this process until the entire string is marked at one-meter intervals. To make tabs, fold 5-cm pieces of masking tape in half around the string, positioning them right before or after each meter mark. Label the tabs, beginning with "0 m" at the knot. Starting with the higher-numbered end, wrap the string around a can to keep it from tangling. To easily wrap and unwrap the string, insert a dowel through both ends of the can.

To measure distances to the nearest meter, one student should hold the zero end of the string while another holds the dowel and unwinds the string. If greater accuracy is desired, a single meter tape can be used together with the string measure. For example, if the length of a tree's shadow is between 24 and 25 meters, a meter tape can be placed along the 24-meter mark of the string, allowing measurement to the nearest centimeter.

Materials
30-50 m of string
Masking tape
Meter tape
Permanent marker
Scissors
Potato chip can or 1 lb. coffee can
Dowel, optional

HARDHATTING IN A GEO-WORLD © 2004 AIMS Education Foundation

Measuring Mass

Balance

Goal: 1 per group

Materials
Ruler
70 cm of string
2 plastic cups
2 paper clips
Tape
Scissors

Tape a 20-cm piece of string to the top center of the ruler (or thread through the center hole and tie). Open the paper clips and tape or hang them on the two ends of the ruler so that the hooks hang below the bottom of the ruler. Tape a 30-cm piece of string to each cup and hang the cups on the paper clips. If the ruler does not balance evenly, add pieces of clay or tape small objects to the side that is high or reposition the top center string to the right or left.

Non-customary masses

Materials
Use objects with uniform mass such as plastic bears or tiles.

Customary masses

Goal: 1 set per group (10 g, 20 g, 50 g, 100 g)

Materials for uniform masses
4 empty film canisters
   (free at most stores that do their own photo developing)
Salt, sand, or BBs
Permanent marker
5-minute epoxy glue, optional
Colored gummed labels, optional
Gram masses or centicubes

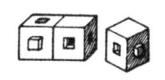

To make a 10-gram canister, place 10 grams on one side of the balance and a film canister and its lid on the other side. Add salt, sand, or BBs until the balance is equalized. If desired, apply epoxy around the top edge to seal the lid in place, then label the canister. Continue the same process to make 20-g, 50-g, and 100-g canisters. (For 100 g, BBs will be needed.) The set can be completed with the addition of commercial 1-g masses or centicubes that are 1 gram.

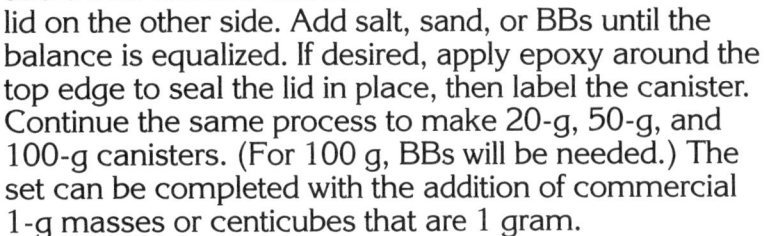

If colored labels are used to color-code mass sets, the labels should be placed on the canister before determining mass.

Materials for non-uniform masses
Rocks, bolts, blocks of wood, old locks, etc.

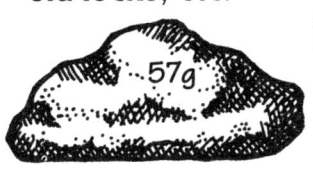

Measure and label the mass of non-uniform objects. These objects then become measuring tools. For example, it might take a 57-gram rock, a 91-gram block of wood and 8 small paper clips (4 grams) to balance a pair of scissors.

HARDHATTING IN A GEO-WORLD © 2004 AIMS Education Foundation

Measuring Volume

Graduated cylinders

Goal: 1 set per group (small, medium, and large jars)

Materials
Small, medium, and large straight-sided jars
Permanent marker
Graduated cylinder or a large syringe (found in pet stores), marked in milliliters

Have students use the graduated cylinder or syringe to measure an amount of water equal to the increment chosen for each jar. Small jars would most likely be calibrated in 10-milliliter increments, larger jars in 25-, 50-, or 100-mL increments. Pour the water into the jar, and carefully mark the level of the water and label the unit on the side of the jar. Repeat this process until you have filled the jar.

Individual or group sets can be stored in shoe boxes, one of several ways to organize and manage these measuring tools.

— adapted from *Student Made Measuring Tools* by Dave Youngs

RULERS LINE UP

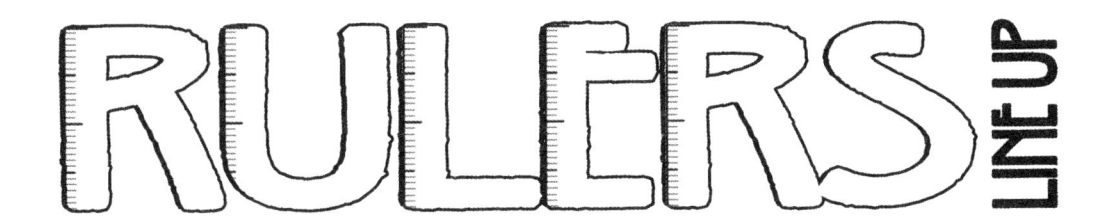

Topic
Linear measurement

Key Question
How can you divide a strip for measuring into ten equal units?

Learning Goals
Students will:
- become familiar with metric units, and
- see the need for using customary units of measure.

Guiding Documents
Project 2061 Benchmark
- *When people care about what is being counted or measured, it is important for them to say what the units are (three degrees Fahrenheit is different from three centimeters, three miles from three miles per hour).*

NRC Standard
- *Tools help scientists make better observations, measurements, and equipment for investigations. They help scientists see, measure, and do things that they could not otherwise see, measure, and do.*

NCTM Standard 2000*
- *Recognize the attributes of length, volume, weight, area, and time*

Math
Linear measurement
  deci-

Integrated Processes
Observing
Comparing and contrasting
Collecting and recording data
Interpreting data
Generalizing

Materials
Construction paper
Lined paper
Overhead transparencies

Background Information
This activity provides students with the opportunity to measure many different objects using a non-customary unit. They will divide their unit into 10 equal parts to gain practice in measuring deci-units. The multiple measuring experiences done here can later be transferred to the use of meter and decimeter measures.

At the conclusion of this activity, students should discover that their non-customary units are not well known to others. This fact makes communication of their measures difficult; therefore, they should begin to see the need to use customary units when trying to communicate the length of objects to others.

Management
1. Cut various lengths of construction paper strips for students to use as measuring devices. Make each strip about 2.5 cm (1 inch) wide and between 15-30 cm (6-12 inches) in length. Each student should have his or her own measuring strip which will be called a roo.
2. The method of dividing the paper strips into 10 equal-sized parts requires a good deal of problem-solving skills—especially with the longer strips. Allow students the time to grapple with finding a solution.
3. Students will need ruled notebook or tablet paper.
4. To help facilitate the procedure for calibrating the measuring strip, make a transparency of a ruled sheet of paper or draw equally spaced lines on the transparency film. Cut your own measuring strip from another sheet of transparency film—a colored transparency adds good contrast.
5. When the lined-paper method is used to calibrate the rulers, those students with longer strips of paper may have difficulties. Encourage them to seek suggestions from other students. Some strategies they may discover are: two sheets of lined paper can be taped together; every other line can be numbered to spread out the scale.
6. Prior to this activity, students should have had multiple experiences using non-customary units of linear measurement such as: footsteps, hand spans, pencil lengths, paper clip lengths, etc.

HARDHATTING IN A GEO-WORLD   37   © 2004 AIMS Education Foundation

Procedure
1. Distribute construction paper strips, one to each student. Inform the students that these strips, roos, will be the rulers with which they will measure several different objects.

☐ = 1 roo

2. Invite the students to find something in the classroom that is one roo long. Have them record the object and its length, using the name of their unit.
3. Allow time for students to find and record things on *Chart A* that measure two, three, and five roos. If necessary, remind them to label their measures.
4. Ask the students to find something that is one-half a roo. After the students have located the objects, have them share their strategies for "knowing" how to determine one-half a ruler length.
5. Inform the students that you want them to be able to measure in tenths. Ask them how they could equally divide their rulers into 10 equal parts. (Many will say to fold it in half, in half again, and in half once more before they discover that this strategy will not work for finding tenths.) Elicit a variety of strategies, allowing time to try them.
6. Tell the students that you are going to share a method that will help them divide their rulers into equal parts. Place the lined transparency on the overhead projector. Ask the students how many parts they need in order to divide their rulers into tenths. [10] Turn the lined transparency so that the lines run vertically. Demonstrate for the students how to number the lines from zero to 10.

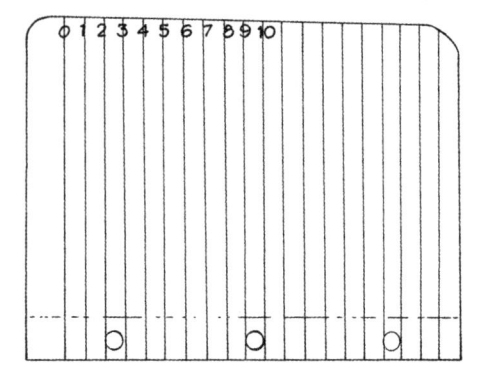

7. Take the transparent ruler and place the upper left corner on the zero line. (The ruler will be longer than the 10 numbered lines.) Ask the students what they should do since the ruler won't fit between the zero and the 10. (If say to cut it off, tell them that they can't alter the length of their ruler.)

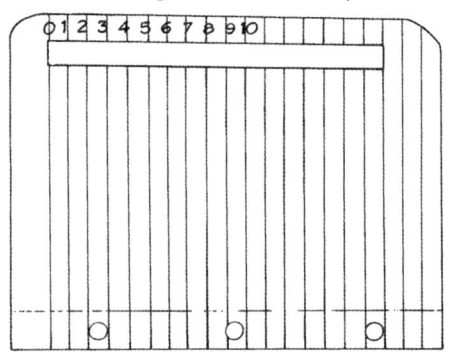

8. Demonstrate to the students how to hold the upper left corner of the ruler on the zero line and to pivot the ruler until the upper right corner is on the 10 line. (It may help to have students highlight line 10.)

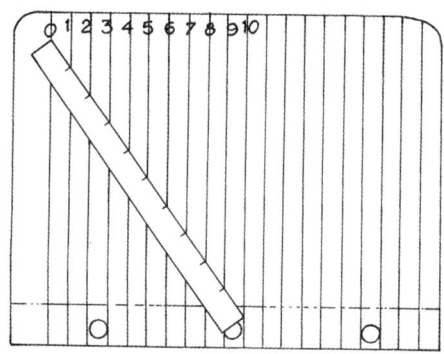

9. Show them how to make little marks where the lines on the transparency intersect the top edge of their ruler.

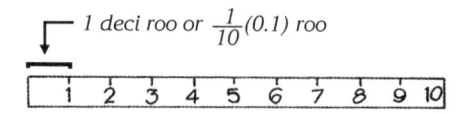

1 deci roo or $\frac{1}{10}$ (0.1) roo

10. Once all marks are in place, have the students number the marks.
11. Tell them that "deci" is the prefix for tenth, so if something is three deci-roos in length, they would need to measure to the three-tenth's mark.
12. Now have the students find objects in the room for the measures listed in *Chart B*.
13. Finally, ask students to record the measure of the objects listed in *Chart C* to the nearest deci-roo.

Connecting Learning
1. How does your roo compare with the person's next to you?
2. Is it important that each of your deci-roos be the same size? Explain.
3. Why is it important to label our measures?
4. Compare the length of your pencil using your roo with the length another person determines for that same pencil using their roo. Are they the same or different? Explain any differences in the measure.
5. What do you think the reaction of a clerk at a home improvement store would be if you called and asked for a certain measure of lumber using your roo units? How could this type of problem (using units that are not well known) be avoided? [Use units that are well known such as meters and feet.]
6. What other problems are there with using roos? [Our measured lengths aren't the same because our roos aren't the same length.]
7. Deci means one-tenth so you divided your ruler into ten equal parts. Centi is the prefix that means one-one hundredth. What do you think you would have to do to your ruler in order to measure centi-units? [divide it into 100 equal-sized parts] How could you do that? [Each of the one-tenth parts needs to be marked into 10 equal-sized parts.] Does it seem reasonable to use the lined paper? Explain. [No, because the lines are too far apart. We would probably just have to estimate the one hundredths.] (Note: Prefixes for the metric system that end in the letter i, such as deci—1/10, centi—1/100, and milli—1/1000, are all fractional parts of the metric unit.)

Extensions
1. If students are able, have them record their measures in units using decimals, fractions, and deci-units. For example: 2.4 roos, 2 4/10 roos, 24 deci-roos.
2. Have students use the lined paper to divide their measuring strips into other fractional parts such as fourths, eighths, etc.
3. Have students approximate the centi-roos on their rulers. Have them measure and record to the nearest centi-roo.
4. Do *Metric Scavenger Hunt* from the AIMS publication *Math + Science, A Solution*.

Curriculum Correlation
Literature
Myller, Rolf. *How Big is a Foot?* Dell Publishing. New York. 1990.

* Reprinted with permission from *Principles and Standards for School Mathematics*, 2000 by the National Council of Teachers of Mathematics. All rights reserved.

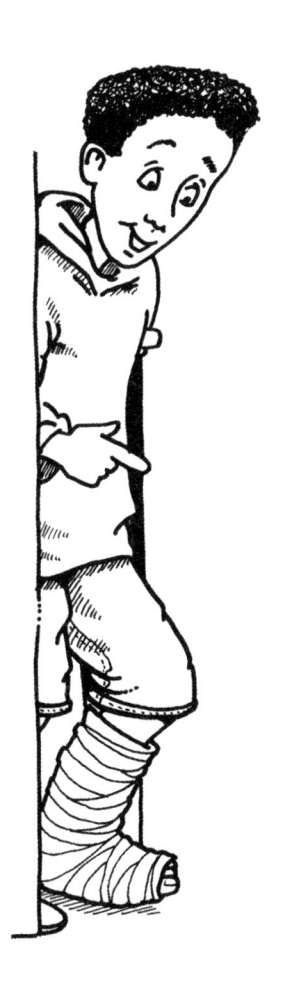

RULERS LINE UP

Key Question

How can you divide a strip for measuring into ten equal units?

Learning Goals

Students will:

- become familiar with metric units, and
- see the need for using customary units of measure.

HARDHATTING IN A GEO-WORLD © 2004 AIMS Education Foundation

RULERS LINE UP

CHART A

Object	Length
	1 roo
	2 ___
	3 ___
	5 ___
	½ ___

CHART B

Object	Length
	1 deci roo
	2 deci ___
	3 deci ___
	4 deci ___
	6 deci ___
	10 deci ___
	12 deci ___
	7/10 deci ___
	11/10 deci ___

CHART C

Object	Length
Your pencil	
This paper	
A shoelace	
Height of door jam	
Hair	

HARDHATTING IN A GEO-WORLD © 2004 AIMS Education Foundation

RULERS LINE UP

CONNECTING LEARNING

1. How does your roo compare with the person's next to you?

2. Is it important that each of your deciroos be the same size? Explain.

3. Why is it important to label our measures?

4. Compare the length of your pencil using your roo with the length another person determines for that same pencil using their roo. Are they the same or different? Explain any differences in the measure.

5. What do you think the reaction of a clerk at a home improvement store would be if you called and asked for a certain measure of lumber using your roo units? How could this type of problem (using units which are not well known) be avoided?

6. What other problems are there with using roos?

7. Deci means one-tenth so you divided your ruler into ten equal parts. Centi is the prefix that means one-one hundredth. What do you think you would have to do to your ruler in order to measure centi-units? How could you do that? Does it seem reasonable to use the lined paper? Explain.

HARDHATTING IN A GEO-WORLD © 2004 AIMS Education Foundation

Links to Length

Topic
Measuring meters

Key Question
How long a paper chain can you make from one piece of paper?

Learning Goals
Students will:
- make the longest chain possible from one piece of paper, and
- measure and graph the results.

Guiding Documents
Project 2061 Benchmarks
- *Length can be thought of as unit lengths joined together, area as a collection of unit squares, and volume as a set of unit cubes.*
- *Spreading data out on a number line helps to see what the extremes are, where they pile up, and where the gaps are. A summary of data includes where the middle is and how much spread is around it.*

NRC Standards
- *Plan and conduct a simple investigation.*
- *Communicate investigations and explanations.*

NCTM Standards 2000*
- *Select and apply appropriate standard units and tools to measure length, area, volume, weight, time, temperature, and the size of angles*
- *Represent data using tables and graphs such as line plots, bar graphs, and line graphs*

Math
Measurement
  linear
Estimation
  rounding
Problem solving
Graphing
Statistics
  range, mode

Integrated Processes
Observing
Collecting and recording data
Comparing and contrasting
Controlling variables

Materials
For each group of two:
  scratch paper
  one 12" x 18" piece of construction paper
  glue or transparent tape
  scissors

For the class:
  meter tapes, sticks, or string
  construction paper strips (see *Management 6*)
  2-3 meters of roving yarn
  20 small paper squares

Background Information
Students are challenged to make the longest paper chain possible from one piece of paper. Those students who plan carefully by identifying the factors that will influence the results will likely make longer chains. One factor is the width of the paper strips. The narrower the strips, the more loops can be made. Another factor is the amount of overlap used when taping or gluing the loops. The most efficient and length-extending method is to bring the ends of the strips together and tape them where they meet. If glue is used, some overlap is necessary. The length of the paper strips is a third factor. Longer strips mean fewer loops and less length lost due to overlap or the intertwining of the loops.

A line plot is used to graphically organize and represent data in this activity. This technique was developed in the late 1970s by a statistician named J.W. Tukey. From a line plot, we can find the most frequent values and see the overall distribution of the results in much the same way as is portrayed in a bar graph. It is another tool for organizing and presenting data.

Length in Meters

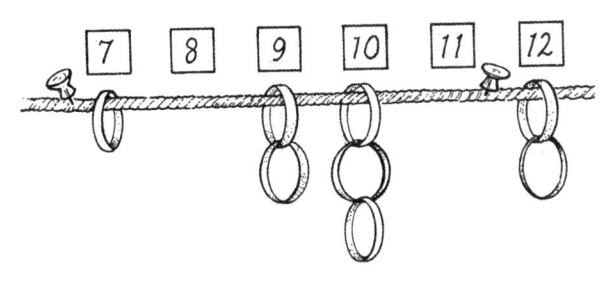

Each loop represents one chain. This line plot shows one 7-m chain, two 9-m chains, and so forth.

HARDHATTING IN A GEO-WORLD © 2004 AIMS Education Foundation

Management

1. Planning, actual construction, and measuring the chain may be spread over two or more days.
2. Students should be paired. Working with a partner allows the sharing of ideas during planning and also speeds the completion of the chain.
3. Use a variety of construction paper colors.
4. A series of colorful student-made meter tapes are ideal for seeing the meter as a repeated unit when measuring the paper chains. One long tape can be used for all measurements. Find directions in *Student-Made Measuring Tools*.
5. Rules you may wish to consider using:
   a. A paper chain is defined as a series of complete loops or links. All of each paper strip must form a single loop.
   b. Tape cannot be used to extend loops. The strips of paper must either meet or overlap each other.
   c. Chains must be sturdy enough to be stretched and not break during measuring. Decide whether students will be allowed to flatten loops in half to extend their length.
6. Prepare the class line plot (graph). A variety of materials could be used. Yarn and paper squares are one suggestion. String the yarn parallel to the floor and at a height where it is visible and accessible to students, perhaps along a bulletin board or chalkboard. Attach the paper squares at even intervals. They will become the meter labels once the range has been determined. Cut paper strips, about 9" x 3/4", one for each pair of students.
7. The finished chains may be used to create a colorful room sculpture. Try intertwining chains as well as stringing them along the walls and hanging them from the ceiling.

Procedure

1. Ask the *Key Question*. Give student pairs scratch paper and tape or glue with which to experiment. Allow sufficient time for students to test strategies using the scratch paper and develop a plan of action.
2. Distribute the activity sheet and have students record their plan.
3. Give each pair the piece of construction paper and have them construct a chain.
4. Instruct students to measure the length of their chain. They can round to the nearest meter or centimeter and record the results as, for example, 10 m, 37 cm or 10.37 meters.
5. Bring the class together and ask them how the range for the class could be found. (They might suggest taking a quick poll.) Finding the range will determine the numbering for the line plot. For example, if no chain is longer than 20 meters, the paper squares could be numbered 1 to 20. If the shortest is 8 m and the longest 25 m, the numbering might begin at 7 and stop at 26.
6. Give each group a paper strip and have them form a loop around the yarn and under the number showing the length of their chain rounded to the nearest meter. If more than one group has the same length, their loops should be attached to the preceding loops in a chain-like fashion. (See the illustration in *Background Information*.)
7. Direct students to record the class line plot on their activity sheet by numbering the line and drawing the loops.
8. Have students identify the range and mode. Hold a concluding discussion.

Connecting Learning

1. What things did you think about when you were making your plans? [the width of the paper strips, the length of the strips, the amount of overlap or how to make the loop without wasting length, etc.]
2. Share your plan.
3. What plan was the most successful? What was good about the plan?
4. If you were to try this again, what would you do differently?
5. If you had a 15-sheet package of the same-sized paper you used today, would your chain go around the perimeter of your classroom? How many times? Would your chain cover the length of a football field? (A football field is 110 meters from end zone to end zone or 91 meters from goal line to goal line.)

---

* Reprinted with permission from *Principles and Standards for School Mathematics*, 2000 by the National Council of Teachers of Mathematics. All rights reserved.

Links to Length

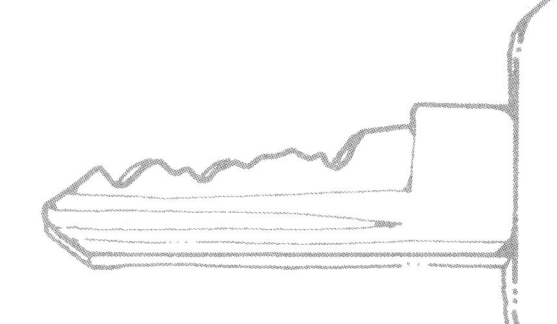

Key Question

How long a paper chain can you make from one piece of paper?

Learning Goals

Students will:

- make the longest chain possible from one piece of paper, and
- measure and graph the results.

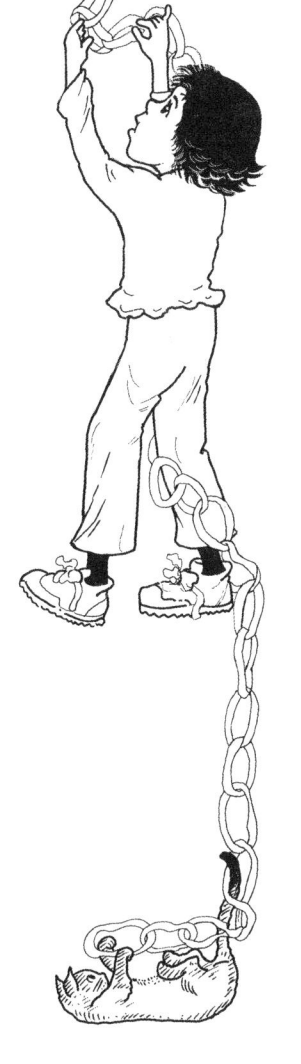

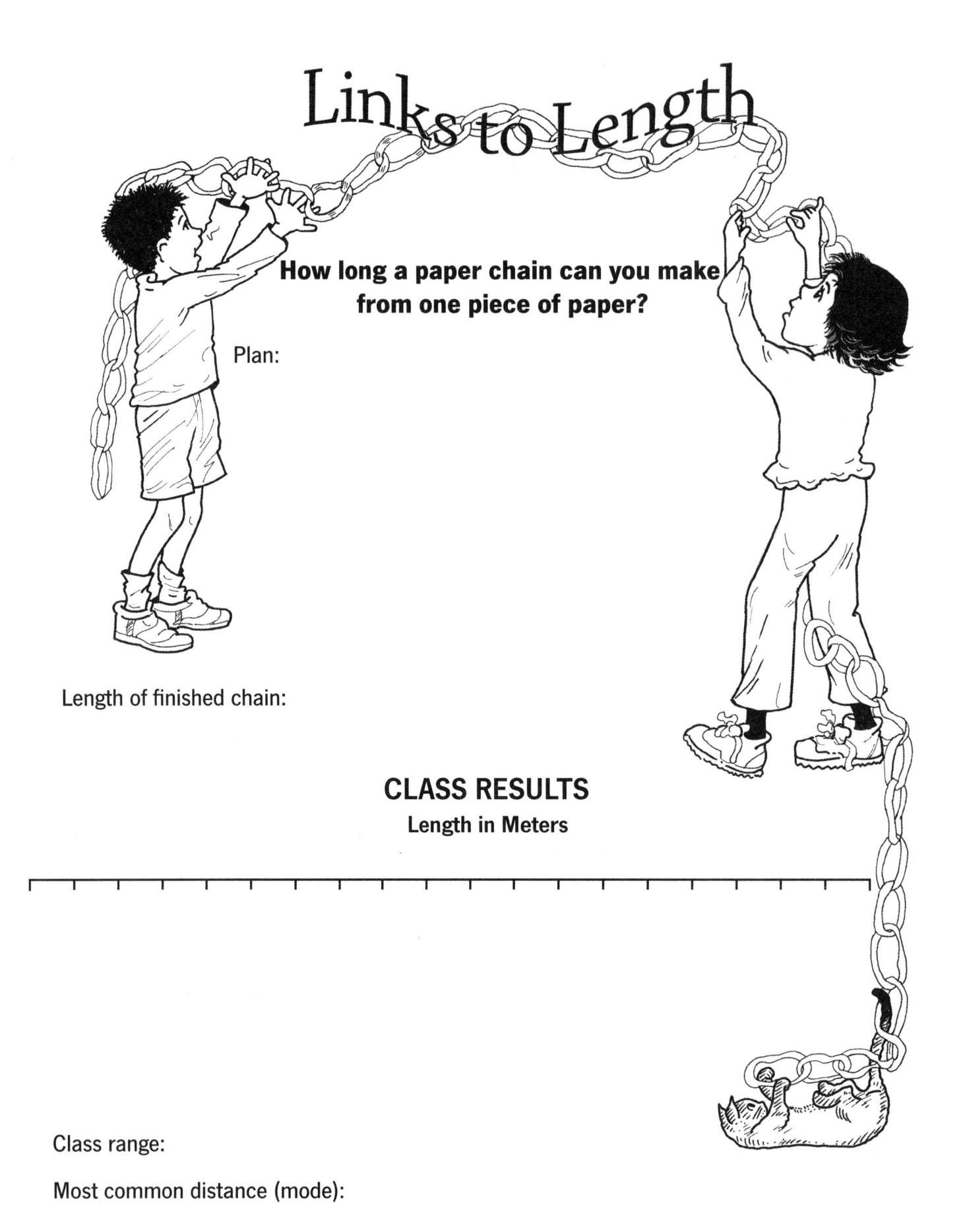

Links to Length

How long a paper chain can you make from one piece of paper?

Plan:

Length of finished chain:

CLASS RESULTS
Length in Meters

Class range:

Most common distance (mode):

Links to Length

1. What things did you think about when you were making your plans?

2. Share your plan.

3. What plan was the most successful? What was good about the plan?

4. If you were to try this again, what would you do differently?

5. If you had a 15-sheet package of the same-sized paper you used today, would your chain go around the perimeter of your classroom? How many times? Would your chain cover the length of a football field? (A football field is 110 meters from end zone to end zone or 91 meters from goal line to goal line.)

Are You A Square?

Topic
Linear measurement/human body

Key Question
How does your height compare with your arm span?

Learning Goals
Students will:
- compare height and arm span, and
- study and interpret class data.

Guiding Documents

Project 2061 Benchmarks
- *Measuring instruments can be used to gather accurate information for making scientific comparisons of objects and events and for designing and constructing things that will work properly.*
- *Measurements are always likely to give slightly different numbers, even if what is being measured stays the same.*

NRC Standards
- *Employ simple equipment and tools to gather data and extend the senses.*
- *Communicate investigations and explanations.*

NCTM Standards 2000*
- *Select and apply appropriate standard units and tools to measure length, area, volume, weight, time, temperature, and the size of angles*
- *Represent data using tables and graphs such as line plots, bar graphs, and line graphs*

Math
Estimation
Measurement
  length
Bar graph

Science
Life science
  human body

Integrated Processes
Observing
Collecting and recording data
Comparing and contrasting
Classifying
Interpreting data

Materials
Meter sticks or tapes (see *Management 2*)
3-column class bar graph (see *Management 1*)
Transparency of *Class Results* page
Sticky notes, 3 colors
Adding machine tape

Background Information
Measurement is an important means of obtaining data. As students measure each other, then double-check those measurements, they should begin to realize that measurements are never exactly the same even though the same thing is being measured. Some discrepancy may be due to careless measuring, but it is also due to the very nature of measurement. Measurement is always an estimation because the increments can always be divided into smaller units. Measuring to the nearest half centimeter is more precise than to the nearest centimeter. Measuring to the nearest millimeter is more precise than to the nearest half centimeter. A millimeter can be divided into still smaller segments.

Because measurement is never exact, it becomes rather clear that a "square" should not just be defined as someone who measures *exactly* the same number of centimeters in height as in arm span. The challenge is to arrive at an acceptable range for the definition of a square.

Several fairly consistent ratios can be found by measuring the human body: head circumference to height, femur to height, etc. In the big picture, the height and the arm span for human beings are nearly the same. Of course there are exceptions to this generalization. A key to your results will be in how a square is defined. Two examples from the world of athletics are:
- Shawn Bradley of the NBA has a height of 7 feet, 6 inches and an arm span of 7 feet, 6 inches.
- Michael Gross, the West German Olympic gold medal swimmer of the 1980s, is 6 feet, 7 inches tall and has an arm span of 7 feet, 4 5/8 inches. (A definite exception to the rule!)

HARDHATTING IN A GEO-WORLD   © 2004 AIMS Education Foundation

Management
1. Prepare a 3-column class bar graph, labeling the columns *square*, *tall rectangle*, and *far-reaching rectangle*. Use a different color sticky note for each body type. Make a matching color key.

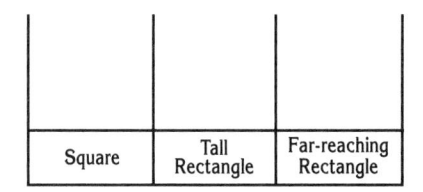

Square	Tall Rectangle	Far-reaching Rectangle

2. For each measuring area, tape the meter sticks to the wall. You will need two meter sticks taped end-to-end vertically to measure height. Also tape two horizontally at an average student's shoulder height to measure arm span. Set up at least two measuring areas.
3. Use groups of three with two students measuring the third one. Height measurement should be taken with shoes off. Place a ruler at the top of the head and parallel to the floor to read the measurement more accurately. The arm span measurement should be taken with the back against the wall and arms outstretched along the meter sticks at shoulder height. Measure from fingertip to fingertip.
4. Allow 45 minutes or more for this activity.
5. To illustrate how a bar graph can be turned into a circle graph, remove the sticky notes and attach them, edge to edge, to a strip of adding machine tape. Placing a piece of paper (at least 24" square) under it, pull the paper strip into a circle. Trace the circle and draw a radius at each place where a new color of sticky notes begins.

Procedure
1. Ask the *Key Question* and give students the activity sheet.
2. Have students estimate and record their height and arm span.
3. Determine whether to round measurements to the nearest millimeter, half centimeter, or centimeter. Discuss how to measure accurately (see *Management 3*).
4. Have groups of three measure each other's height and arm span. They should check their results by measuring more than one time.
5. After recording their measurements, students should complete the L-shaped bar graph and record their height and arm span on the *Class Results* transparency.

6. Put the transparency on the overhead projector. Ask questions such as:
   a. What do you notice?
   b. Who has the greatest difference?
   c. Which students are square?
   d. How much wiggle room (margin for error) do we need? How will we define a square—exactly the same measurements, up to 1 cm difference, up to 2 cm difference? Why? [possible errors in measurement, rounding process]
   e. What does providing wiggle room do to our results? Do we now have more or less squares than at first?
7. Have students compare their results to the three illustrations and record whether they are a square, tall rectangle, or far-reaching rectangle. Using the data on the transparency, have the class also decide whether each person is a square, tall, rectangle, or far-reaching rectangle.
8. Instruct students to place a colored sticky note in the correct column of the class bar graph.
9. Discuss the results. Follow up by showing them how the data can be illustrated with a circle graph (see *Management 5*).

Connecting Learning
1. Are you a square? What information do you need in order to draw that conclusion?
2. What is meant by being a "perfect" square? Were there any perfect squares in the class?
3. Why did we decide to let our squares be less than perfect?
4. Which shape is most common? Which shape is most rare? How could you tell?
5. What conclusions can you draw from this activity?
6. What are you wondering about now? (Would another class have the same results? Does age affect results? Would family members all be the same?)
7. How could you find the answer?

Extension
Measure another class or adults at school. Record and compare.

Curriculum Correlation
Literacy
Write to athletes, actors, etc. and ask them for their height and arm span.

Home Link
Have students measure their family. Does the same shape run in the family?

* Reprinted with permission from *Principles and Standards for School Mathematics*, 2000 by the National Council of Teachers of Mathematics. All rights reserved.

Are You A Square?

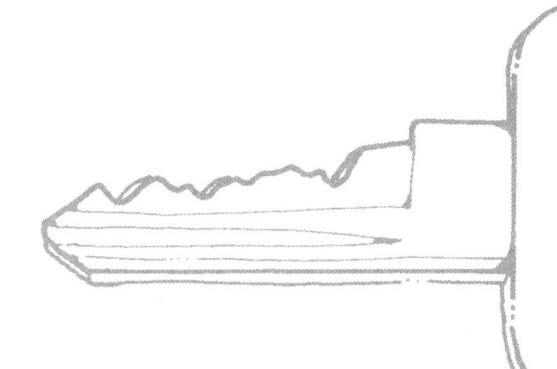

Key Question

How does your height compare with your arm span?

Learning Goals

Students will:

- compare height and arm span, and
- study and interpret class data.

HARDHATTING IN A GEO-WORLD

Are You A Square?

How does your height compare with your arm span?

	Height (cm)	Arm Span (cm)
Estimate	_____	_____
Measure	_____	_____

Graph your height and arm span.

height = arm span
Square

height > arm span
Tall Rectangle

height < arm span
Far-reaching rectangle

Which one are you?

HARDHATTING IN A GEO-WORLD   51   © 2004 AIMS Education Foundation

Are You A Square?

Class Results

	Name	Height	Arm Span	Square	Tall rectangle	Far-reaching rectangle
1.						
2.						
3.						
4.						
5.						
6.						
7.						
8.						
9.						
10.						
11.						
12.						
13.						
14.						
15.						
16.						
17.						
18.						
19.						
20.						
21.						
22.						
23.						
24.						
25.						
26.						
27.						
28.						
29.						
30.						
31.						
32.						

HARDHATTING IN A GEO-WORLD © 2004 AIMS Education Foundation

Are You A Square?

1. Are you a square? What information do you need in order to draw that conclusion?

2. What is meant by being a "perfect" square? Were there any perfect squares in the class?

3. Why did we decide to let our squares be less than perfect?

4. Which shape is most common? Which shape is most rare? How could you tell?

5. What conclusions can you draw from this activity?

6. What are you wondering about now? (Would another class have the same results? Does age affect results? Would family members all be the same?)

7. How could you find the answer?

Bear Facts

Topic
Linear measurement

Key Question
How do you compare to your bear?

Learning Goal
Students will use length measurement to compare themselves to their teddy bears.

Guiding Documents
Project 2061 Benchmarks
- *Measuring instruments can be used to gather accurate information for making scientific comparisons of objects and events and for designing and constructing things that will work properly.*
- *Add, subtract, multiply, and divide whole numbers mentally, on paper, and with a calculator.*

NRC Standard
- *Employ simple equipment and tools to gather data and extend the senses.*

NCTM Standards 2000*
- *Select and apply appropriate standard units and tools to measure length, area, volume, weight, time, temperature, and the size of angles*
- *Develop fluency in adding, subtracting, multiplying, and dividing whole numbers*

Math
Estimation
Measurement
  length
Whole number computation
  subtraction

Integrated Processes
Observing
Collecting and recording data
Comparing and contrasting

Materials
Teddy bears (see *Management 1*)
Metric measuring tapes (see *Management 2*)

Background Information
This activity provides an appealing context for students to practice linear measurement skills, measuring both straight lengths as well as *around* objects.

Management
1. Before doing the activity, ask students to bring a well-loved teddy bear from home. You may wish to have an extra teddy bear available for students who have no bear and wish to adopt one for the day. Or students may work in pairs and share a bear.
2. In preparation for measuring, have students color, cut out, and assemble the measuring tape found in the back of this book. To measure student height, vertically attach two measuring tapes to the wall.
3. Students need to work in pairs to measure each other. This works nicely, too, if they are sharing a bear.

Procedure
1. Announce a "bring your bear to school" day.
2. Have students record the bear's name and picture on the first activity page.
3. Discuss how each measurement will be done. Height should be taken without shoes. Arm length is defined as the shoulder joint to the finger tip, running along the outer part of the arm. Leg length extends from the hip joint (bend a leg to find it) to the sole of the foot, along the outer part of the leg. The head should be measured at its widest point.
4. Instruct students to measure themselves and their bears with the tape and record this data in the table.
5. Have students compare measurements by finding the differences.
6. Discuss the results. Have students answer the two questions at the bottom of the page.
7. Distribute the second page. Tell students to draw pictures of themselves and their bears.
8. Have students compare characteristics beyond size by asking how they are alike and different from their bears. Encourage them to record at least three responses to each question.

Connecting Learning
1. How much taller are you than your bear?
2. How many bears tall are you?
3. Which measurement—height, arm, leg, or head—is closest to being the same as the bear?
4. In what other ways can we compare ourselves to our bears?

Extensions
1. Arrange bears in a line or a train from small to tall or light to dark ...etc.
2. Make a collection of bear facts for a class Big Book that tells about bears at school.
3. Choose one bear and measure each child's height in bears. Make a bear graph of the results.

* Reprinted with permission from *Principles and Standards for School Mathematics*, 2000 by the National Council of Teachers of Mathematics. All rights reserved.

BEAR FACTS

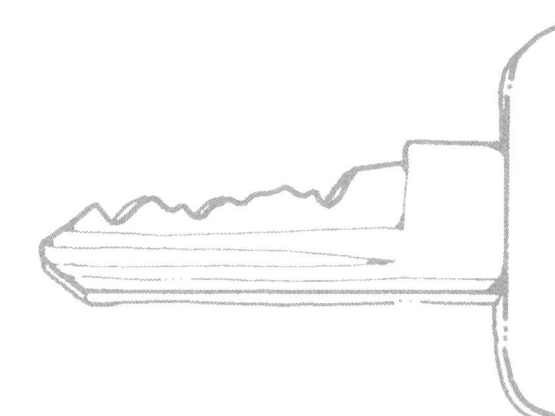

Key Question

How do you compare to your bear?

Learning Goal

Students will:

- use length measurement to compare themselves to their teddy bears.

BEAR FACTS

How do you compare to your teddy bear?

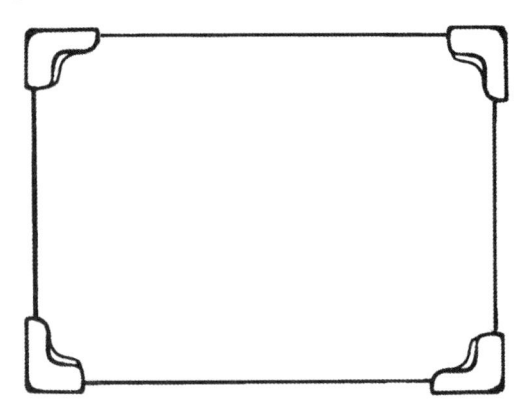

Bear's Name _____

	Height (cm)	Arm Length (cm)	Leg Length (cm)	Around Head (cm)
Me				
Bear				
Difference				

How many bears tall are you?

What makes your bear special?

HARDHATTING IN A GEO-WORLD 56 © 2004 AIMS Education Foundation

BEAR FACTS

How do you compare to your bear?

ME   BEAR

How are you like your bear?

How are you different from your bear?

HARDHATTING IN A GEO-WORLD © 2004 AIMS Education Foundation

BEAR FACTS

1. How much taller are you than your bear?

2. How many bears tall are you?

3. Which measurement—height, arm, leg, or head—is closest to being the same as the bear?

4. In what other ways can we compare ourselves to our bears?

Cups 'n' Stuff

Topic
Measuring mass

Key Question
If volume is equal, how does the mass compare?

Learning Goals
Students will:
- find the mass of cups filled with different materials, and
- order the cups from heaviest to lightest.

Guiding Documents
Project 2061 Benchmarks
- When people care about what is being counted or measured, it is important for them to say what the units are (three degrees Fahrenheit is different from three centimeters, three miles from three miles per hour).
- Use numerical data in describing and comparing objects and events.

NRC Standards
- Objects have many observable properties, including size, weight, shape, color, temperature, and the ability to react with other substances. Those properties can be measured using tools, such as rulers, balances, and thermometers.
- Tools help scientists make better observations, measurements, and equipment for investigations. They help scientists see, measure, and do things that they could not otherwise see, measure, and do.

NCTM Standards 2000*
- Understand such attributes as length, area, weight, volume, and size of angle and select the appropriate type of unit for measuring each attribute
- Select and use benchmarks to estimate measurements

Math
Measurement
  mass, customary
  volume, non-customary
Ordering
Estimation
  rounding
Graphing

Integrated Processes
Observing
Collecting and recording data
Comparing and contrasting
Controlling variables
Drawing conclusions

Materials
For each group:
  1/2 cup each of five different materials
    (see *Management 2*)
  5 small plastic cups, about 3 oz each
  balance
  gram masses
  glue
  crayons or colored pencils

Background Information
  The intent of this activity is to give students experience in determining mass. If some of the materials have nearly the same mass, students may also realize that using a balance to measure is a more accurate way of comparing than by lifting the cups.
  The underlying reason for the differences in mass is the varying densities of the materials. Students at this level are not yet ready to formally deal with density, but they are given an experience that will lead to the presentation of that concept when they are developmentally ready.

Management
1. Organize the class into groups of four or five. The groups should work together but each group member will benefit by completing an activity sheet.
2. Each group will need about 1/2 cup or 125 mL each of five different materials. Food suggestions: pinto or navy beans, popcorn kernels, rice, salt or sugar, barley, macaroni, lentils, cereal, cornmeal, etc. Other suggestions: sand, vermiculite, pet food, dirt, etc.
3. It is helpful to have students fill the cups for their group earlier in the day. Schedule the rotation so the measuring area will not be crowded. To control volume, cups of the same size should be filled to the brim with each material and leveled by sliding an index card across the top.

HARDHATTING IN A GEO-WORLD © 2004 AIMS Education Foundation

Procedure

1. Have each group retrieve their filled cups. Review the fact that they have measured equal volumes of five different materials. Then ask the *Key Question*.
2. Explain that each group will order the materials from heaviest to lighest by two methods, lifting and measuring. For students who have difficulty finding a systematic strategy for lifting the cups, suggest the following method:
   a. Compare all five cups by lifting two at a time, one in each hand. Find the heaviest and the lightest and place them on a flat surface about a ruler's length apart. Each group member should have a turn lifting the cups and then the group should agree on order placement.
   b. Compare the three that are left. Find the heaviest and the lightest and place them in the appropriate order between the other two cups.
   c. Ask, "Where does the one that is left belong?"
3. To record the order obtained by lifting, have students glue a few pieces of the material in each of the cups on the activity sheet.
4. Direct groups to gather a balance and gram masses, then measure and record the mass of each cup of material. You may wish to have students include an estimate of mass as well as actual mass.

Material	Estimated Mass	Actual Mass

5. Have students graph the results in order from heaviest to lightest. Remind them to label the graph and give it a title.
6. To record their measuring order in another way, instruct students to glue a sample of each material in the second set of cups on the activity sheet.
7. Revisit the *Key Question* as you discuss the results.

Connecting Learning

1. How did your lifting results compare with your measuring results? [Lifting is less accurate than measuring so, for materials with nearly the same mass, there may be differences in order.]
2. Explain how you know the volume was the same.
3. Use *greater than* and *less than* to compare the masses of the cups. (Example: macaroni < salt)
4. What is your conclusion about mass when volume is equal? [It is likely mass will not be equal.]
5. What other materials could be measured in the same way? What do you predict about the results?
6. How could we plan to test the mass of liquids with the same volume? (Students will need to find a solution for potential spilling problems. One solution: Use one cup as the volume measurer, drawing a line somewhere below the brim. Measure each liquid up to this line, then pour it into another cup. Rinse and dry before each new liquid is measured.)
7. What are you wondering now?

Extension

Do the activity again using liquids, as designed by students in response to *Discussion 5*.

* Reprinted with permission from *Principles and Standards for School Mathematics*, 2000 by the National Council of Teachers of Mathematics. All rights reserved.

Cups 'n' Stuff

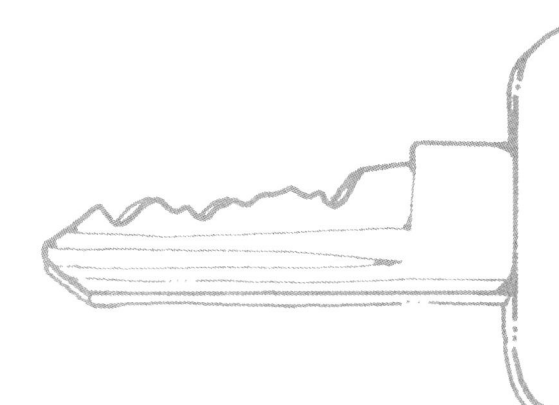

Key Question

If volume is equal, how does the mass compare?

Learning Goals

Students will:

- find the mass of cups filled with different materials, and
- order the cups from heaviest to lightest.

Cups 'n' Stuff

If volume is equal, how does the mass compare?

heaviest					lightest

By Lifting

heaviest					lightest

By Measuring

Measure and record the mass of each material.

Put in order from heaviest to lightest on the graph.

HARDHATTING IN A GEO-WORLD   62   © 2004 AIMS Education Foundation

Cups 'n' Stuff

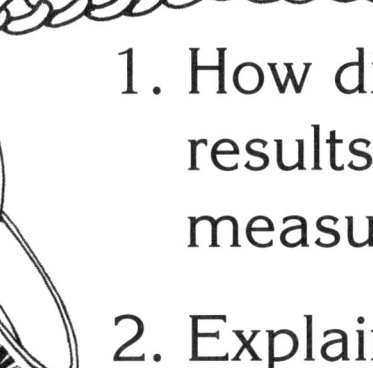

1. How did your lifting results compare with your measuring results?

2. Explain how you know the volume was the same.

3. Use *greater than* and *less than* to compare the masses of the cups.

4. What is your conclusion about mass when volume is equal?

5. What other materials could be measured in the same way? What do you predict about the results?

6. How could we plan to test the mass of liquids with the same volume?

7. What are you wondering now?

Peddle the Metal

Topic
Measuring mass

Key Question
How much money do I need in order to buy your jewelry collection?

Learning Goals
Students will:
- make pasta jewelry, and
- determine cost by measuring mass.

Guiding Documents
Project 2061 Benchmarks
- When people care about what is being counted or measured, it is important for them to say what the units are (three degrees Fahrenheit is different from three centimeters, three miles from three miles per hour).
- Judge whether measurements and computations of quantities such as length, area, volume, weight, or time are reasonable in a familiar context by comparing them to typical values.

NRC Standard
- Objects have many observable properties, including size, weight, shape, color, temperature, and the ability to react with other substances. Those properties can be measured using tools, such as rulers, balances, and thermometers.

NCTM Standards 2000*
- Select and apply appropriate standard units and tools to measure length, area, volume, weight, time, temperature, and the size of angles
- Develop fluency in adding, subtracting, multiplying, and dividing whole numbers

Math
Measurement
 mass
Money
Whole number operations
 multiplication

Integrated Processes
Observing
Collecting and recording data
Comparing and contrasting
Applying

Materials
String
Scissors
Pasta (see *Management 2*)
Balances
Gram masses
Calculators
Paper for price tags and display

Background Information
Many items we buy are priced per unit measure. At the grocery store, fresh fruit and vegetables as well as meat are sold by the pound. Medication, in the form of pills, is labeled by number of milligrams. Gold is a commodity that is also priced this way; in some markets, even gold jewelry is sold by the gram.

Students have the opportunity to simulate the practice of pricing according to mass through a combination art project/measuring experience. In addition to challenging their creativity, they are able to apply gram measurement in a realistic way. They also work with our money system and must determine whether the prices they calculated are reasonable for the numbers used.

Management
1. Divide the class into groups of four or five.
2. Gather three or more kinds of pasta that can be strung such as salad macaroni, elbow macaroni, and mostaccioli.
3. To make colored pasta simulating beads or emeralds, rubies, and sapphires, pour about 1/4 cup rubbing alcohol and several drops of food coloring into a jar with a lid. Add the pasta, close the jar, and shake to coat. Spread the pasta on several layers of newspaper to dry. Do this in a well-ventilated area.
4. Determine the cost per gram based on the ability level of your students. The simplest approach is to set the cost in whole dollars (such as $3 per gram.) If you want students to deal with decimals, choose a cost below one dollar (such as $.85) for easier computation, or above one dollar (such as $4.70) for more challenging computation. Another option is to check the newspaper in order to use the current price of gold. (One ounce of gold equals 28 grams.)
5. Have the materials available in a designated area. One or two members from each group can gather what they need.

HARDHATTING IN A GEO-WORLD © 2004 AIMS Education Foundation

6. You can control the activity in one of three ways: a) limit the time for creating the jewelry, b) limit the number of jewelry pieces that can be made, or c) limit the number of grams of pasta used by each group, say 100 grams. (For this last option, use the *Challenge* shown on the easel on the activity page in place of the *Key Question*.)

Procedure
1. Tell students that each group will be designing and displaying a jewelry collection. They can make matching sets of jewelry or individual pieces. Explain that jewelry is sometimes priced by the number of grams it measures. The cost of their jewelry will be set at ____ per gram.
2. Distribute the activity page and have students write the cost per gram.
3. Have one or two persons from each group gather some starting supplies of string and pasta. Instruct the groups to make their jewelry collections.
4. Direct each group to get a balance and gram masses. Have them find and record the mass of each piece of jewelry on the activity page, then compute its price according to the cost set per gram.

Example

Jewelry	Mass	Price (at $3 per gram)
necklace 1	16 g	$48.00
necklace 2	22 g	$66.00
bracelet	7 g	$21.00

5. Have groups set up a jewelry display complete with price tags including the number of grams.
6. Organize a way for the groups to go window shopping among the displays.
7. Discuss the results.

Connecting Learning
1. How many things can you think of that are sold by mass or weight? (fresh fruits and vegetables, pills, gold jewelry, etc.)
2. How did you know your price calculations were correct? (Students should explain their reasoning for why their answers made sense. For example, if 16 grams were calculated at $.85 per gram and the answer was $18, students should reason that because $.85 is below one dollar and there are 16 grams, the answer has to be less than $16. For the same problem, if they obtained an answer of $136, they should reason that it could not possibly be that much and question whether they put the decimal in the right place. Correct answer: $13.60)
3. Is your jewelry collection designed for wealthy people, people without very much money, or is there something for every budget? Explain.
4. Which one or two pieces of your collection do you consider the most unique?

Extensions
1. Students who are wearing jewelry might find the cost of their pieces using the same cost-per-gram as in the activity.
2. Have students figure the price of their jewelry based on the current price of gold. Gold prices are reported by the ounce. One ounce equals 28 grams.
3. Expand the activity into a consumer simulation by having students actually buy and sell their jewelry using play money. They could record their total sales for the day. Optional: include the current sales taxes.
4. For older students, determine a "cost per hour" for labor. Have them calculate the cost of the gold used to make their jewelry and add on the labor cost to determine the selling price. You may also want to attach a percentage value for profit.

* Reprinted with permission from *Principles and Standards for School Mathematics*, 2000 by the National Council of Teachers of Mathematics. All rights reserved.

Peddle the Metal

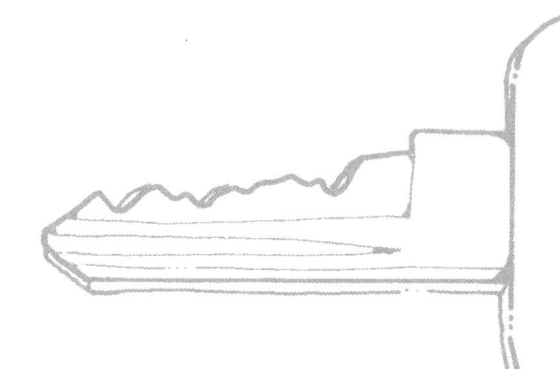

Key Question

How much money do I need in order to buy your jewelry collection?

Learning Goals

Students will:

- make pasta jewelry, and
- determine cost by measuring mass.

Peddle the Metal

Design and make a collection of jewelry to be sold at _____ a gram.

Record the mass and price of each piece.

Make a window display of your jewelry showing number of grams and price.

HARDHATTING IN A GEO-WORLD © 2004 AIMS Education Foundation

Peddle the Metal

1. How many things can you think of that are sold by mass or weight?

2. How did you know your price calculations were correct?

3. Is your jewelry collection designed for wealthy people, people without very much money, or is there something for every budget? Explain.

4. Which one or two pieces of your collection do you consider the most unique?

Filling Stations

Topic
Measuring capacity/volume

Key Question
How much will these containers hold?

Learning Goal
Students will compare the capacity of various containers.

Guiding Documents
Project 2061 Benchmarks
- Mathematical ideas can be represented concretely, graphically, and symbolically.
- When people care about what is being counted or measured, it is important for them to say what the units are (three degrees Fahrenheit is different from three centimeters, three miles from three miles per hour).
- Tables and graphs can show how values of one quantity are related to values of another.

NRC Standard
- Tools help scientists make better observations, measurements, and equipment for investigations. They help scientists see, measure, and do things that they could not otherwise see, measure, and do.

*NCTM Standards 2000**
- Understand such attributes as length, area, weight, volume, and size of angle and select the appropriate type of unit for measuring each attribute
- Represent data using tables and graphs such as line plots, bar graphs, and line graphs

Math
Measurement
    capacity/volume
Graphing

Integrated Processes
Observing
Predicting
Collecting and recording data
Comparing and contrasting

Materials
For each group:
    5 different containers (see *Management 2*)
    graduated cylinders
    bucket or plastic tub
    scissors
    12" x 18" construction paper
    transparent tape or glue

Background Information
Both *volume* and *capacity* are measures of three-dimensional spaces. Volume is the amount of **space occupied** by an object or a material in an object. Capacity is the amount of **available space** in an object, how much it holds. In this activity, we are comparing the capacity of various containers by using a graduated cylinder to measure the volume of water each one holds.

A graduated cylinder is generally labeled in milliliters. It takes 1000 milliliters to make one liter. The liter, a basic unit of volume measurement, is defined as 1000th of a cubic meter. It is a little larger than a quart.

Students often have limited experience with three-dimensional measurement, yet it has many useful applications including cooking, buying gasoline, taking liquid medicines, and comparing amounts in beverage containers. This activity is one means of expanding measurement skills.

Management
1. Divide the class into groups of four or five. Although students will work in groups, consider having each student complete an activity sheet and graph.
2. Gather containers of similar size but of different shapes, heights, and widths such as tomato paste cans, tomato sauce cans, tuna cans, cups and jars of various sizes, half-pint milk cartons, etc.
3. For the containers suggested in *Management 2*, 100-, 250-, and 500-milliliter graduated cylinders will be useful. The 100-mL graduated cylinder allows for more precise readings.
4. To measure, set the graduated cylinder on a flat surface and read at eye level. Because the water is attracted to the sides of the graduated cylinder, it will form a concave shape. Remind students to read the water level at the bottom of the meniscus, near the center of the water line.

HARDHATTING IN A GEO-WORLD © 2004 AIMS Education Foundation

5. Each group will need a large water container such as a five-gallon plastic bucket (check the cafeteria) or a plastic tub. It will hold the group's water supply and help contain spills as students pour and measure.
6. For the 3-D graph cylinders, make colored copies of the line page included or use lined writing paper. Each student will need five cylinders.

   Alternatively, students can construct their own 2-D bar graphs labeled *Containers* (horizontal axis) and *Milliliters* (verticle axis).
7. If possible, do this activity outdoors or in a place where water spills are not a problem. A few towels may be needed for mopping spills.

Procedure
1. Have the class assemble into groups. Display one set of containers, introduce the *Key Question*, and distribute the activity page.
2. Instruct students to draw the five containers on their page. The containers may be labeled by description such as "tomato paste" or with letters such as A, B, C, D, and E.
3. Direct groups to write the predicted order of the containers from largest capacity to smallest capacity, using the labels on their drawings.
4. Have each group take their water supply, graduated cylinders, and containers to the designated area. Instruct them to fill each container to the brim with water and carefully pour it into a large graduated cylinder. This should be done over their bucket or tub. For more precise readings, they can then transfer smaller quantities to a 100-mL graduated cylinder. A record should be made, either next to their drawings on the activity page or on a piece of scratch paper.
5. Oversee the clean-up of the workspace and equipment and have students return to the classroom.
6. Distribute the construction paper and lined paper for the graph.
7. Have students determine the numbering of the graph based on their highest milliliter reading and number each of the five lined papers in the same way. Tell them to cut along the line marked zero.

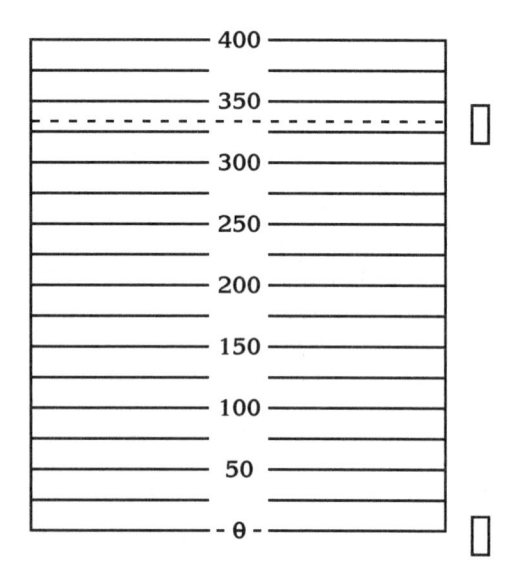

8. For each graph cylinder, direct students to use a straight edge to mark the number of milliliters measured for a container, cut along the mark, roll it into a cylinder (overlapping so lines meet), and tape or glue.
9. Tell students to attach the cylinders to the construction paper, in order from smallest capacity to largest capacity, to form a graph like that shown on the activity page. Either label or tape/glue the picture drawn of the container under each cylinder.

Connecting Learning
1. Why is it important to set the graduated cylinder on a table top before reading the water level? [It prevents the water from being tipped which causes inaccurate readings.]
2. The results of which container most surprised you?
3. How close were your predictions to your results?
4. How much would a container hold if you had to use the entire lined section of your paper?
5. What did you most enjoy about this activity?

Extensions
1. Find a simple recipe and have students use metric measures to make the food.
2. Do the related activity, *All Bottled Up*, found in the AIMS publication *Water, Precious Water*.

Home Link
1. Have students check the liquid measuring cup in their kitchens at home. Most now have two scales, cups and milliliters. About how many milliliters are in one cup?
2. Encourage students to look in their pantries for food containers labeled with number of milliliters. They might record the products and number of milliliters and share this data with the class. What kinds of products are labeled in milliliters?

* Reprinted with permission from *Principles and Standards for School Mathematics*, 2000 by the National Council of Teachers of Mathematics. All rights reserved.

Filling Stations

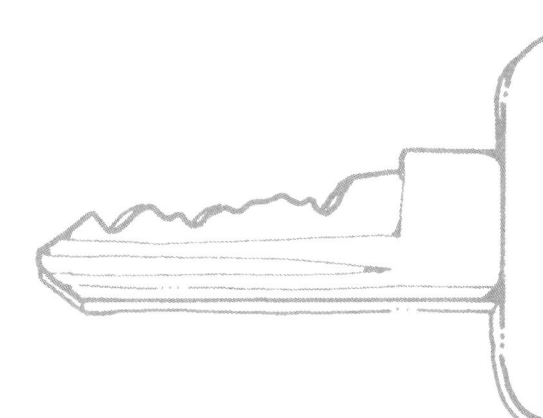

Key Question

How much will these containers hold?

Learning Goal

Students will:

- compare the capacity of various containers.

HARDHATTING IN A GEO-WORLD

FILLING STATIONS

How much will these containers hold?

Draw and name the containers.

Predicted order

Use lined paper rolled into a cylinder to show the number of milliliters for each container. Attach the cylinders to a peice of construction paper in order from smallest capacity to largest.

HARDHATTING IN A GEO-WORLD © 2004 AIMS Education Foundation

HARDHATTING IN A GEO-WORLD 73 © 2004 AIMS Education Foundation

Filling Stations

1. Why is it important to set the graduated cylinder on a table top before reading the water level?

2. The results of which container most surprised you?

3. How close were your predictions to your results?

4. How much would a container hold if you had to use the entire lined section of your paper?

5. What did you most enjoy about this activity?

Pleased as Punch

Topic
Measuring volume

Key Question
Which punch looks and tastes best?

Learning Goals
Students will:
- plan various mixes of punch,
- measure according to their recipes, and
- judge the results.

Guiding Documents
Project 2061 Benchmarks
- When people care about what is being counted or measured, it is important for them to say what the units are (three degrees Fahrenheit is different from three centimeters, three miles from three miles per hour).
- Tables and graphs can show how values of one quantity are related to values of another.
- Measure and mix dry and liquid materials (in the kitchen, garage, or laboratory) in prescribed amounts, exercising reasonable safety.

NRC Standard
- Employ simple equipment and tools to gather data and extend the senses.

NCTM Standards 2000
- Understand such attributes as length, area, weight, volume, and size of angle and select the appropriate type of unit for measuring each attribute
- Represent data using tables and graphs such as line plots, bar graphs, and line graphs

Math
Measurement
 volume
Graphing

Integrated Processes
Observing
Collecting and recording data
Comparing and contrasting
Controlling variables

Materials
For the class:
 3-6 kinds of drink mix (see *Management 2*)

For each group:
 about 15 transparent cups, 6 oz or more
 100-mL graduated cylinder or 9-oz transparent cups with graduated scales
 crayons or colored pencils

Background Information
Volume is a measure of three-dimensional space. It is the amount of space occupied by an object or a material in an object. In this activity, students use graduated cylinders to measure the volume of each kind of punch that is part of their recipes.

The graduated cylinder is generally labeled in milliliters. One thousand milliliters make one liter, a little more than a quart.

Volume measurement is often less familiar to students than linear, mass, and other forms of measurement. Its practical applications extend from cooking and baking to the buying of gasoline and labeling of beverages. Through measuring experience, such as that provided in this activity, students can build their understanding of volume.

Management
1. Have students work in groups of four or five.
2. When purchasing the drink mix, choose flavors with vivid coloring such as grape, cherry, strawberry, orange, and lime. Juice-box size packages are ideal for each group.
3. Be mindful of students who are diabetic or for some other reason should not ingest sugar.
4. Consider doing this activity outdoors, as spills are probable. Have paper towels handy.
5. Caution students not to plan all the recipes at one time because the results of one may lead to a different kind of mix to try.
6. Graduated cylinders should be washed with soap and water before and after this activity.
7. Each group will need one cup for each drink mix flavor (labeled with the flavor), one cup for each of their recipes (labeled A, B, C, and D), and one taste-testing cup for each group member (labeled with each person's name.)

HARDHATTING IN A GEO-WORLD   75   © 2004 AIMS Education Foundation

Procedure
1. "Today you have the chance to create a new punch by mixing any combination of the drinks we have. Your challenge is to find a combination that looks and tastes the best."
2. Distribute the activity sheet and explain that each group first needs to make a recipe or plan. To do this, they should color the *Recipe A* graduated cylinder to represent the mix of punches and amounts they want to use, in increments of 10 and making a total of 100 milliliters. If necessary, they should create a key. It is helpful to write the amounts to be measured next to each color.

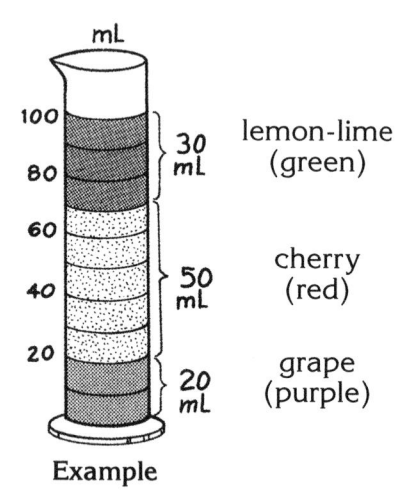

Example

- 30 mL lemon-lime (green)
- 50 mL cherry (red)
- 20 mL grape (purple)

3. Have each group gather a graduated cylinder, eight to 11 transparent cups, and a labeled cup of each kind of punch in the designated area. Have students label four cups A, B, C, and D. The remaining cups will be used for taste tests. Each student should label one cup with his/her name.
4. To avoid over- and under-measuring errors, have students measure one kind of punch at a time, following their recipe. Using the above example, they would first pour 20 mL of grape punch into the graduated cylinder, adding or pouring back a little until they get the right amount. When the amount is correct, it should be poured into the transparent cup labeled A. Next, they should measure 50 mL of cherry punch and add it to cup A, followed by 30 mL of lemon-lime punch.
5. Instruct each group to use their activity sheet to draw or describe the punch they created.
6. Have students follow the same procedure for *Recipe B*, then *Recipe C*, and finally *Recipe D*.
7. Direct each group to conduct a taste test and declare their winning recipe (A, B, C, or D). Have them give their punch an interesting name.

Connecting Learning
1. What problems, if any, did you have mixing the punch?
2. What, to your group, is a good-looking punch?
3. How did your group come to agree on the winning punch?
4. Why did you need to keep a record of the measurements in the different recipes?.
5. Compare the recipes of other groups. Can you find some that are similar to yours? What recipes did other groups have that you would like to try?
6. What are you wondering now?

Extension
Have students make a simple recipe using metric measurements.

Curriculum Correlation
Art
Have students design a packet for their favorite punch.

Literacy
Have students write an advertisement for their favorite punch. They can include a jingle or a song. Let the students make a video of themselves performing the advertisement.

* Reprinted with permission from *Principles and Standards for School Mathematics,* 2000 by the National Council of Teachers of Mathematics. All rights reserved.

Pleased as Punch

Key Question

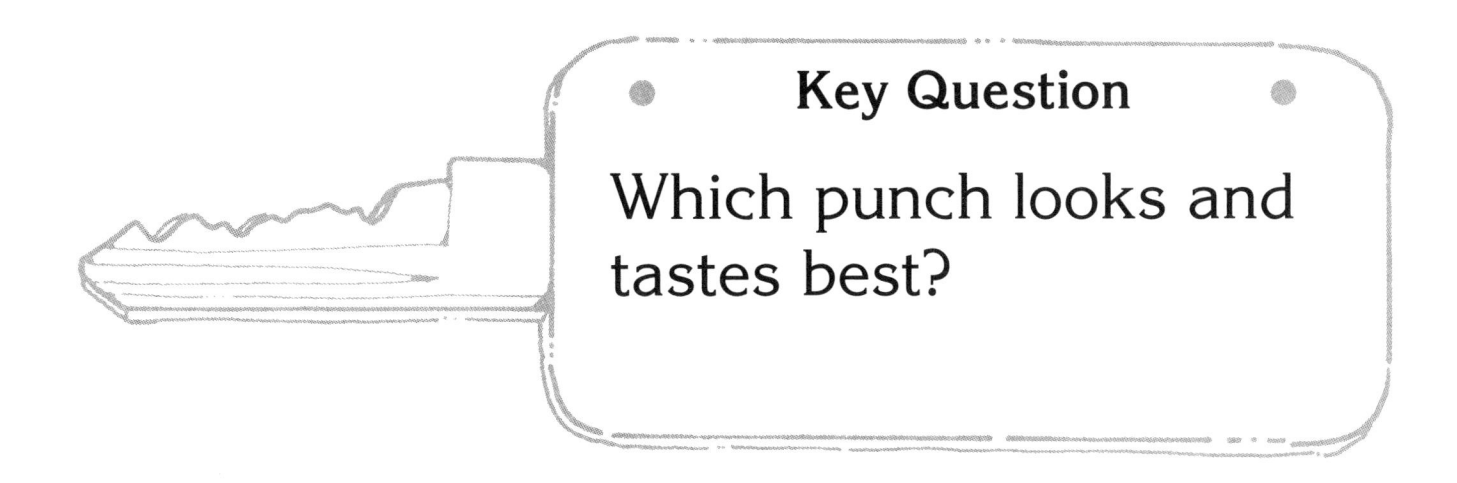

Which punch looks and tastes best?

Learning Goals

Students will:

- plan various mixes of punch,
- measure according to their recipes, and
- judge the results.

Pleased as PUNCH

Make four 100-mL samples of punch mixed in any combination.

Which punch looks and tastes the best?

| Recipe A Results | Recipe B Results | Recipe C Results | Recipe D Results |
|---|---|---|---|//
| | | | |

... and the winning recipe is _____!

Name of punch:

Pleased as Punch

Graduated Scales

Tape the graduated scale to the outside of a 9 oz transparent cup. Cover completely with transparent tape.

HARDHATTING IN A GEO-WORLD © 2004 AIMS Education Foundation

Pleased as Punch

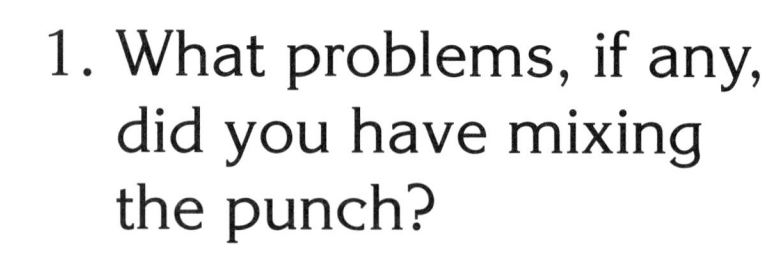

Connecting Learning

1. What problems, if any, did you have mixing the punch?

2. What, to your group, is a good-looking punch?

3. How did your group come to agree on the winning punch?

4. Why did you need to keep a record of the measurements in the different recipes?

5. Compare the recipes of other groups. Can you find some that are similar to yours? What recipes did other groups have that you would like to try?

6. What are you wondering now?

Minute Minders

Topic
Time: one minute

Key Question
How reliable are your one-minute timekeepers?

Learning Goals
Students will:
- perform various tasks for one minute, and
- repeat the tasks to test how reliable they are as timekeepers.

Guiding Documents
Project 2061 Benchmarks
- *Keep records of their investigations and observations and not change the records later.*
- *Offer reasons for their findings and consider reasons suggested by others.*
- *Use numerical data in describing and comparing objects and events.*

NRC Standard
- *Employ simple equipment and tools to gather data and extend the senses.*

*NCTM Standards 2000**
- *Select and use benchmarks to estimate measurements*
- *Collect data using observations, surveys, and experiments*

Math
Measurement
 time
Estimation

Integrated Processes
Observing
Collecting and recording data
Comparing and contrasting
Drawing conclusions

Materials
Clock or other time-piece with a second hand
Ball that bounces
Scratch paper

Background Information
The purpose for this activity is to help students become aware of the length of a minute as well as to realize the need for measuring time based on an even rate.

Students will find that the tasks performed are not altogether reliable as timekeepers. One of the reasons is that variables are not controlled. Sometimes a ball bounces higher, sometimes lower. Your hand might get tired as you write your name continually. Jumping jacks may not be done at an even rate. You can decrease or increase the number of breaths you take a minute. Even if a conscious effort were made to control variables, there would be discrepancies. It is not easy for a person to control the variations with which they write their name or bounce a ball. This points to the need for a consistent measure of time—clocks and watches. The tasks are interesting, though, as an approximation of a minute.

It is tempting for students to manipulate or change the results in order to reach their estimate or to have their second timing match one minute. This is a good opportunity to point out that a scientist's goal is to find the truth. Fudging the facts will not give you the truth.

Management
1. Students should work with a partner.
2. Students are given four repetitious tasks to perform. You may want to devise a short list of possibilities from which students will choose the fifth task. It should require little equipment and be able to be performed by an individual as opposed to a team.
3. Students will first be timed estimating when a minute has passed. Then they will be timed twice during each task, once while performing the task for one minute and then while duplicating the previous number of repetitions of the task.
4. The tasks may be done in any order. In fact, you may want to organize pairs so some start with bouncing a ball outside while others are writing their name inside. Caution: They should count breaths either before doing tasks involving exercise or after they are sufficiently recovered from them.
5. Emphasize that this is not a competition with others in the class. If it were, variables would have to be controlled. Instead, the purpose is for students to gain a sense of the length of a minute.

HARDHATTING IN A GEO-WORLD © 2004 AIMS Education Foundation

Procedure
1. *Preliminary timing:* "How long is a minute? Explain that each person will have a chance to guess when a minute has passed. Pair students and have one of each pair face away from the clock. Give an audible "start" and have the person facing away from the clock tell their partner when they think a minute has passed by softly saying, "Stop" or "Now." The partner can then tell them how many seconds have passed.
2. Have the partners trade places and repeat, with the other partner estimating when a minute has passed. "Who estimated less than a minute? Who estimated more than a minute? How close were you?"
3. *First timing of task:* Give students the activity sheet and explain the procedure. Have students choose and write the fifth task in the blank.
4. Instruct students to estimate how many times they can do the first task. Their partner should then time them, giving soft, audible "start" and "stop" signals. Emphasize that they should make an honest effort at the task and not try to match their estimate.
5. Have partners switch roles, with the one who just finished performing the task now becoming the timer.
6. *Second timing of task:* Inform students that they will again perform the same task. This time the performer will use his/her previous results to try to determine when one minute has elapsed while facing away from the clock. (If a name was written 17 times in a minute the first time, the name should again be written 17 times.)
7. The partner should give the "start" signal. The person doing the task should count his/her repetitions and say "stop" when the number needed is reached. The partner will then tell how much time elapsed.
8. Repeat *Steps 4* through *7* for each task.
9. Bring the class back together to discuss and draw conclusions raised by the *Key Question*, "How reliable are your one-minute timekeepers?"

Connecting Learning
1. How did the time of your second trial compare with one minute?
2. How reliable are these activities as timekeepers? [They are all right for estimating time but not reliable as a more precise measure of time.] Why do you think this is so? [It is difficult to perform activities at an even rate.]
3. What surprised you about the results?
4. What did you learn about a minute?
5. Even though the length of a minute doesn't change, it sometimes *seems* longer or shorter. When does a minute seem like a long time? (traveling to a vacation spot, waiting for cookies to bake, etc.) When does it seem like a short time? (playing with friends, watching a favorite TV show, etc.)
6. When would it be important not to waste minutes? (when you have a time limit to finish a test or chores, running a race, getting ready to go to a movie, etc.)

Curriculum Correlation
Literacy
Have students explore the meaning of folk sayings and idiomatic expressions about time. They might also illustrate the sayings as Fred Gwynne does in *A Chocolate Moose for Dinner* and *The King Who Rained*, both published in 1988 by Simon and Schuster, Books for Young Readers. Some examples:
   A stitch in time saves nine.
   Lost time is never found again.
   Time flies when you're having fun.
   Time waits for no man.
   In the nick of time
   Time on your hands
   Killing time
   The time of your life
   From time to time
   Behind time
   Big time
   Give a hard time
   In less than no time
   In a minute
   Once upon a time

* Reprinted with permission from *Principles and Standards for School Mathematics,* 2000 by the National Council of Teachers of Mathematics. All rights reserved.

Minute Minders

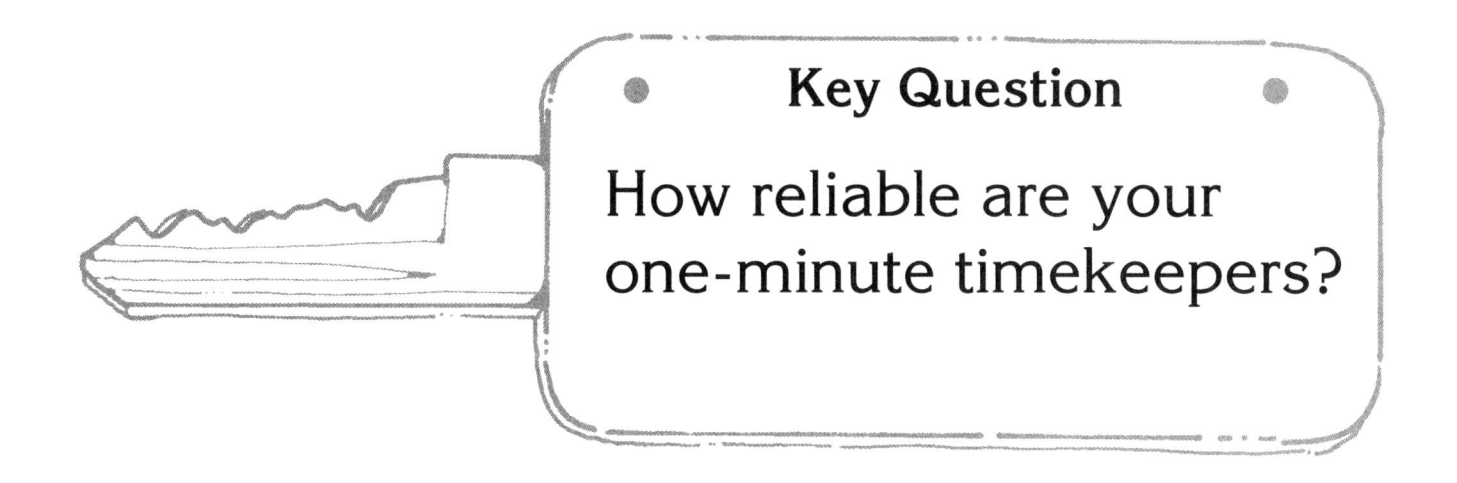

Key Question

How reliable are your one-minute timekeepers?

Learning Goals

Students will:

- perform various tasks for one minute, and
- repeat the tasks to test how reliable they are as timekeepers.

Minute Minders

Partners _____

Estimate how many times you can do each activity in one minute. Try it and record the results. Repeat each activity the number of times you counted and have your partner time how long it takes.

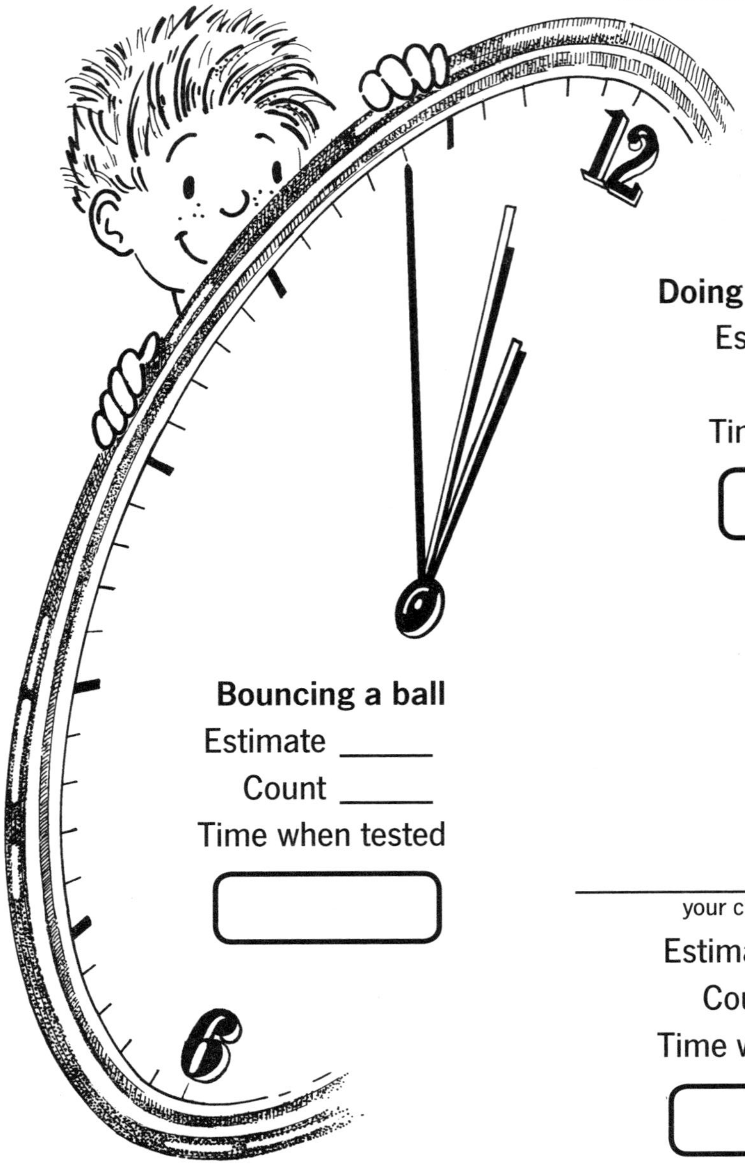

Doing jumping jacks
Estimate _____
Count _____
Time when tested

Bouncing a ball
Estimate _____
Count _____
Time when tested

your choice _____
Estimate _____
Count _____
Time when tested

Writing your name
Estimate _____
Count _____
Time when tested

Taking breaths
Estimate _____
Count _____
Time when tested

How good are these activities as time keepers?

HARDHATTING IN A GEO-WORLD   84   © 2004 AIMS Education Foundation

Minute Minders

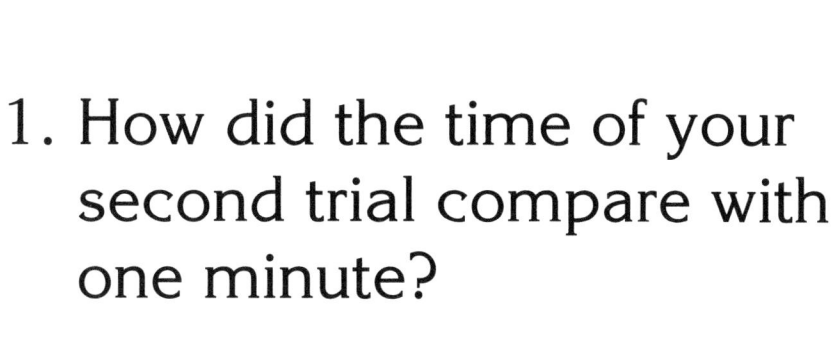

1. How did the time of your second trial compare with one minute?

2. How reliable are these activities as timekeepers? Why do you think this is so?

3. What surprised you about the results?

4. What did you learn about a minute?

5. Even though the length of a minute doesn't change, it sometimes *seems* longer or shorter. When does a minute seem like a long time? When does it seem like a short time?

6. When would it be important not to waste minutes?

From Wedges to Wangles

Topic
Non-customary angle measurement

Key Question
How many wangles fit in these angles?

Learning Goal
Students will use wedge-shaped units (wangles) to measure angles found in the real world and in triangles made on geoboards.

Guiding Documents
Project 2061 Benchmarks
- Mathematical ideas can be represented concretely, graphically, and symbolically.
- Numbers and shapes—and operations on them—help to describe and predict things about the world around us.
- Use numerical data in describing and comparing objects and events.

NRC Standard
- Employ simple equipment and tools to gather data and extend the senses.

NCTM Standards 2000
- Understand such attributes as length, area, weight, volume, and size of angle and select the appropriate type of unit for measuring each attribute
- Measure with multiple copies of units of the same size, such as paper clips laid end to end
- Use geometric models to solve problems in other areas of mathematics, such as number and measurement

Math
Measurement
  angles
Geometry and spatial sense

Integrated Processes
Observing
Sorting
Collecting and recording data
Comparing and contrasting

Materials
For each group of two:
  scissors
  transparent tape
  geoboard
  rubber band

Background Information
Angles are found outdoors, indoors, almost everywhere we look. They are an integral part of many two-dimensional and three-dimensional shapes we encounter, from the painted lines of a basketball court to the silhouette of jagged mountains.

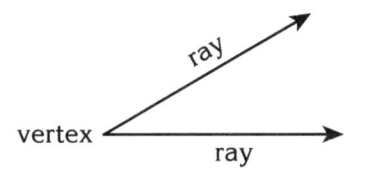

An angle is made of two rays with a common endpoint (vertex). The measure of an angle can be taken by subdividing it into uniform wedge-shaped units. This is a very different kind of measurement than the linear scale on a ruler, a thermometer, or a graduated cylinder frequently used by students. Just as the centimeter is a repeated unit in length measurement, the wedge is a repeated unit in angle measurement.

Since this wedge-shaped unit is largely unfamiliar to both students and adults, understanding begins by using multiple copies laid edge to edge to measure the openness of the rays. This experience lays the foundation for understanding protractors and, eventually, more advanced mathematics.

The wedge-shaped unit used in this activity has been nicknamed a wangle. Since cutting and placement will not be precise, measurements will be approximate.

Management
1. Organize the class into groups of two.
2. Each group will need 12 wangles. Copy the wangle page on card stock.
3. Folding the wangles on the dotted line, though optional, allows students to easily move them to the angle being measured. Fine position adjustments can then be made by sliding.

Procedure
Part One
1. Ask students to name some of the measuring units they have used such as meters, grams, and degrees Celsius. Explain that they will be introduced to a very different kind of measuring unit shaped like a wedge. They will discover into how many of these wedge-shaped units (wangles) an angle can be divided. If necessary, review the terms *ray* and *vertex*.

2. Distribute the card stock wangles, one "quarter-circle" (six wangles) per student. Instruct students to cut and fold as indicated. 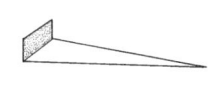
3. Distribute the two picture pages. Direct students to use their wangles to measure the angles formed by the bold lines (rays). Have them record the number of wangles and kind of angle.

Part Two
1. Ask, "What might make it easier to measure with the wangles?" [connect them]
2. Give each student a strip of transparent tape about 9 cm (3.5 in) long. Have them attach their six wangles, without overlapping or gapping, to form a "quarter-circle" protractor.
3. Show the class how the wangle edges can be folded when measuring.
4. Distribute geoboards, rubber bands, and the geoboard page.
5. Have students follow the directions on the page, including some problem solving at the bottom.
6. Discuss the results.

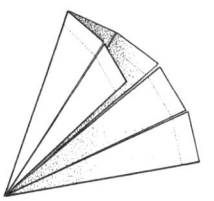

Connecting Learning
Part One
1. How many wangles fit in each picture? What kind of angle is each?
2. Look at each of the pictures again. What other angles in them could be measured?
3. How easy or difficult was it to measure with the wangle units? (Someone may suggest it would be easier if they were all connected, a good lead-in to *Part Two*.)

Part Two
1. How precise were your angle measurements? [Sometimes the wangles fit perfectly, like in square corners. Other times—when the wangles were a little greater than or less than the angle—measurements were approximate.]
2. On your geoboard, make a triangle with angles that can't easily be measured with wangles. What would help you measure these angles? [a tool with smaller units]
3. Look at the 6-3-3 triangle you made. How does it compare to the other 6-3-3 triangle on the page? [It is congruent (same shape/same size) or it is similar (same shape, different size).]
4. Check the sum of the angles in each triangle. What is your conclusion? [The sum of the angles of triangles is 12 wangles.]
5. What objects, inside or outside of the classroom, can you find that can be measured in wangles?
6. How many wangles wide would the door have to be open in order to walk in or out?

Extension
On geoboards, have the class explore the angles of quadrilaterals with challenges such as:
- Make a quadrilateral whose angles all measure 6 wangles. [any rectangle or square]
- Make a quadrilateral with two 4-wangle angles and two 8-wangle angles. [certain trapezoids or parallelograms]
- Make a trapezoid with two 5-wangle angles and two 7-wangle angles.
- What is the sum of the angles of quadrilaterals? [24 wangles]

Curriculum Correlation
Art
Draw a real-world object incorporating a specific angle such as those on the picture pages.

* Reprinted with permission from *Principles and Standards for School Mathematics,* 2000 by the National Council of Teachers of Mathematics. All rights reserved.

Key

First Picture Page (in wangles)
Acute angles *Right angle*
2 gazelle 6 sailboat
3 tripod
4 snowflake

Obtuse angles *Straight angle*
8 bridge, stairs 12 window
10 chair

Second Picture Page
Acute angles *Right angle*
1 scissors 6 bookshelf
2 ferris wheel
4 leaf *Obtuse angle*
5 slide, tree 9 roof

Geoboard Page

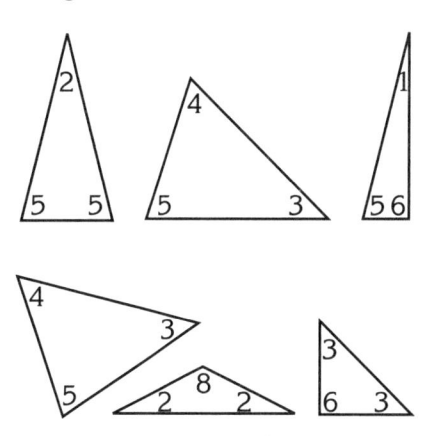

From Wedges to Wangles

Key Question

How many wangles fit in these angles?

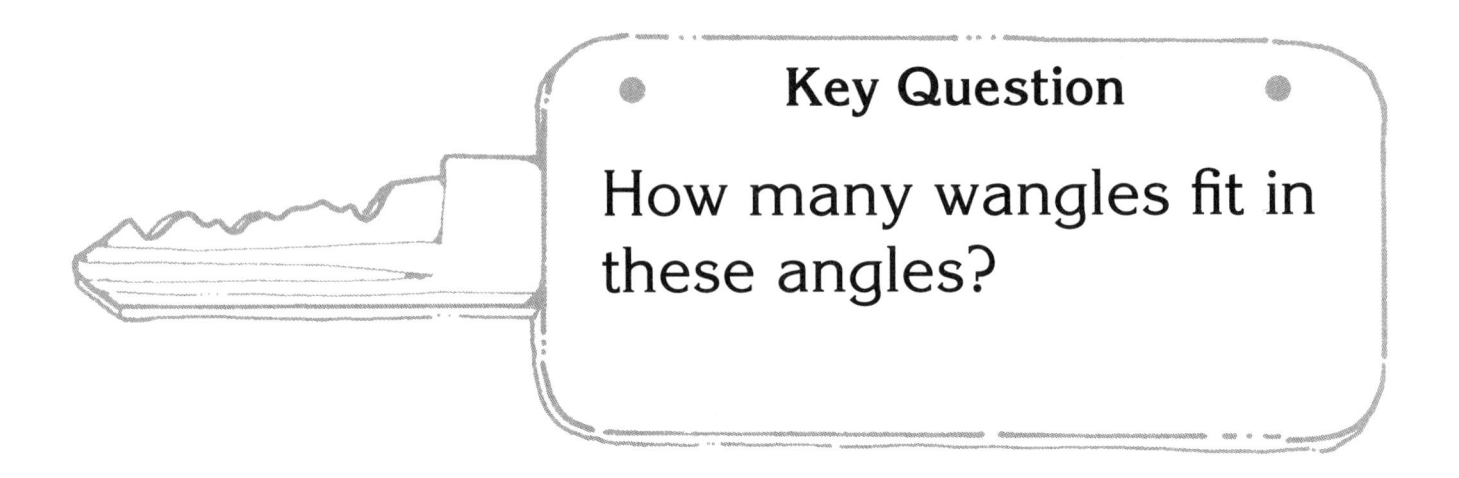

Learning Goal

Students will:

- use wedge-shaped units (wangles) to measure angles found in the real world and in triangles made on geoboards.

From Wedges to Wangles

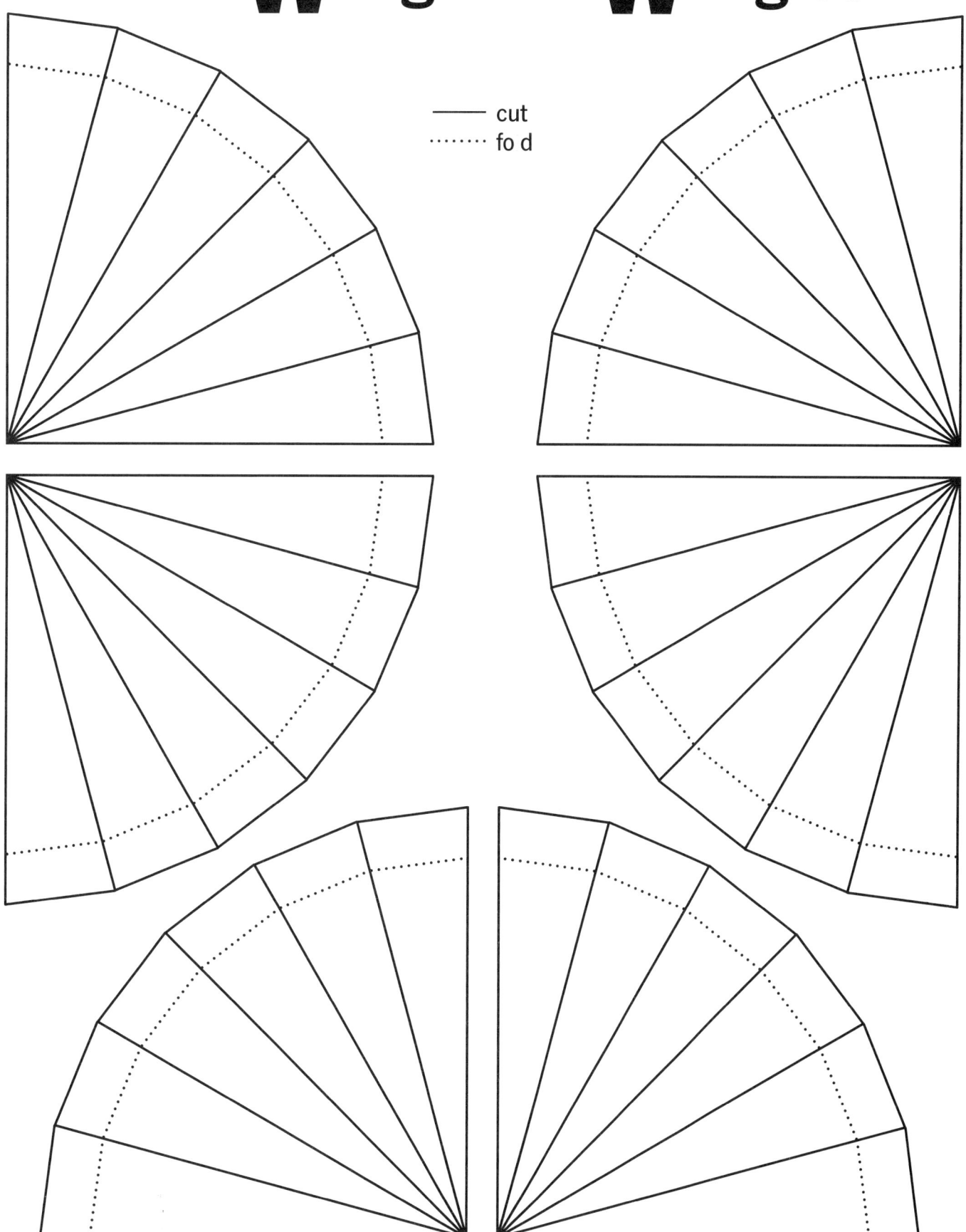

— cut
······· fold

HARDHATTING IN A GEO-WORLD 89 © 2004 AIMS Education Foundation

From Wedges to Wangles

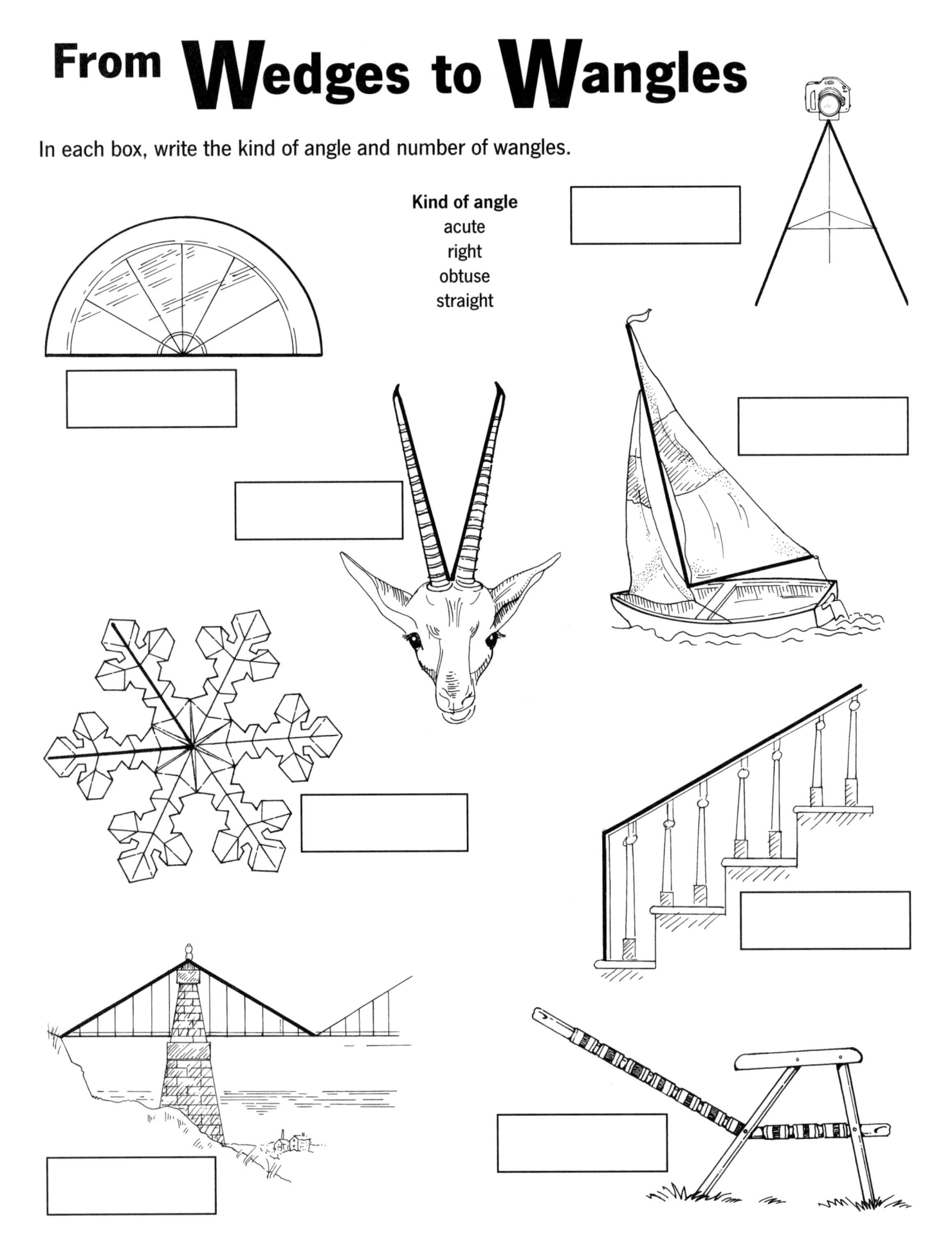

In each box, write the kind of angle and number of wangles.

Kind of angle
acute
right
obtuse
straight

HARDHATTING IN A GEO-WORLD 90 © 2004 AIMS Education Foundation

From Wedges to Wangles

In each box, write the kind of angle and number of wangles.

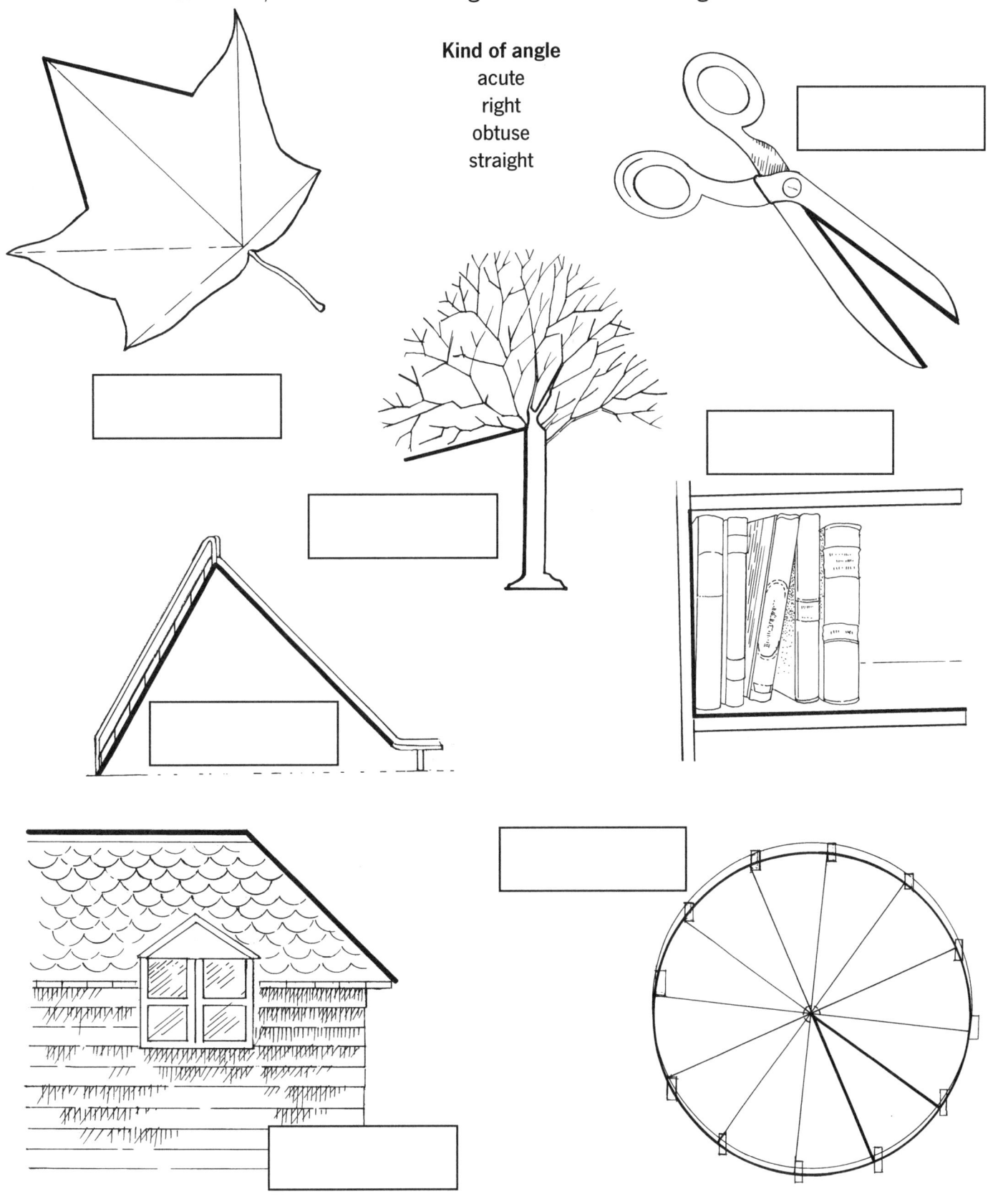

Kind of angle
acute
right
obtuse
straight

From Wedges to Wangles

Tape your wangles together so they look like this.

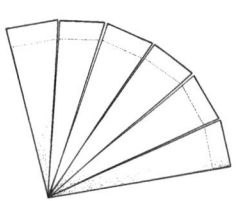

How many wangles are in each angle?

Copy each of these triangles on your geoboard. Measure and record the number of wangles in each angle. The first one has been started for you.

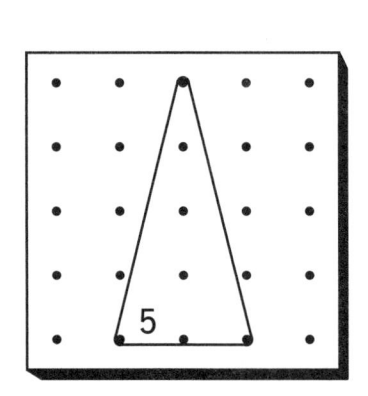

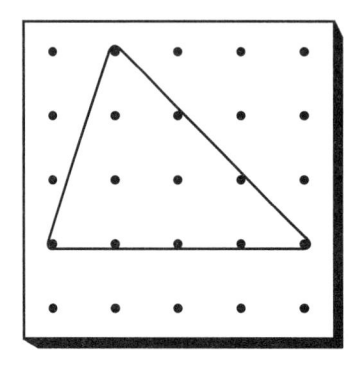

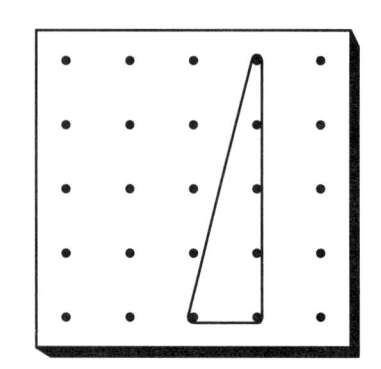

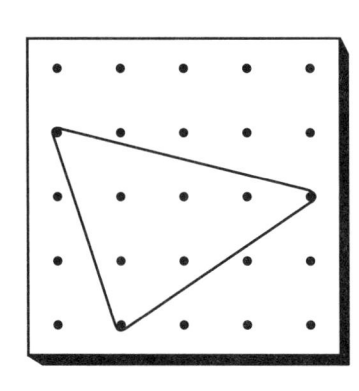

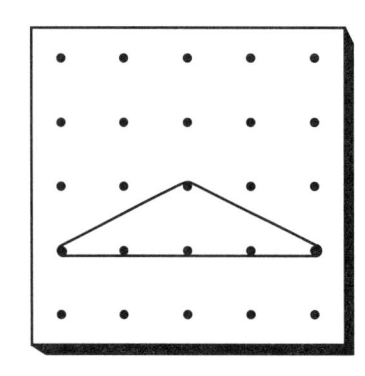

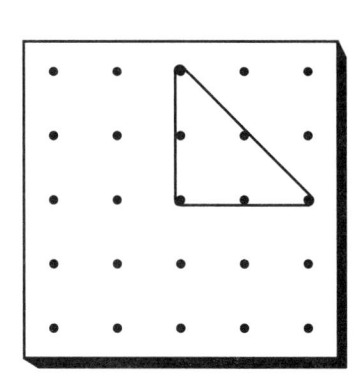

One of the triangles above has angles measuring 6 wangles, 3 wangles, and 3 wangles. Make another 6-3-3 triangle and draw it.

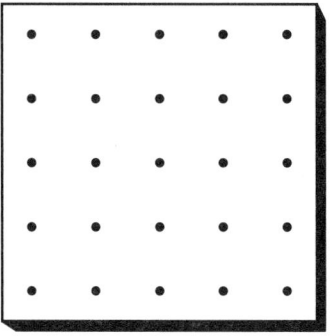

HARDHATTING IN A GEO-WORLD

From Wedges to Wangles

Part One
1. How many wangles fit in each picture? What kind of angle is each?
2. Look at each of the pictures again. What other angles in them could be measured?
3. How easy or difficult was it to measure with the wangle units?

Part Two
1. How precise were your angle measurements?
2. On your geoboard, make a triangle with angles that can't easily be measured with wangles. What would help you measure these angles?
3. Look at the 6-3-3 triangle you made. How does it compare to the other 6-3-3 triangle on the page?
4. Check the sum of the angles in each triangle. What is your conclusion?
5. What objects, inside or outside of the classroom, can you find that can be measured in wangles?
6. How many wangles wide would the door have to be open in order to walk in or out?

Waxed Wangles

Topic
Non-customary angle measurement

Key Question
How can we make a tool to measure angles?

Learning Goals
Students will
- fold a non-customary protractor, and
- use it to construct a clock and measure angles formed by the clock's hands

Guiding Documents
Project 2061 Benchmarks
- *Measuring instruments can be used to gather accurate information for making scientific comparisons of objects and events and for designing and constructing things that will work properly.*
- *Make sketches to aid in explaining procedures or ideas.*
- *Use numerical data in describing and comparing objects and events.*

NCTM Standards 2000
- *Understand such attributes as length, area, weight, volume, and size of angle and select the appropriate type of unit for measuring each attribute*
- *Explore congruence and similarity*
- *Use geometric models to solve problems in other areas of mathematics, such as number and measurement*

Math
Measurement
   angles
   time: reading a clock
Geometry and spatial sense
   symmetry
   congruence
Problem solving

Integrated Processes
Observing
Collecting and recording data
Comparing and contrasting
Applying

Materials
For each student:
   5-inch to 6-inch square of waxed paper or patty paper (see *Management 1*)
   scissors
   1 paper plate or clock circle, 6" to 7"
   2 clock hands
   1 paper fastener

Background Information
This activity continues the exploration of angle measure begun in *From Wedges to Wangles*. An angle can be measured by counting the number of uniform wedge-shaped units into which it can be divided. These non-customary units have been nicknamed wangles. After working with individual wangles, students develop a need for a measuring tool with multiple wangles. They have already experienced taping the wangles together in the previous activity. Now they will be introduced to a translucent and circular wangle protractor.

The clock, being a circle whose hands form an angle, is a natural context for using a protractor. The circle is divided into congruent wedge-shaped units (sectors) whose edges are radii. On a real clock, we only see small ticks, tiny bits of the radii. *Both the clock and the protractor are, in essence, scales derived from wedge-shaped units.*

Reading a clock, part of the study of time measurement, assumes an understanding of its structure. In the cycle of a 24-hour day, one rotation about the Earth's axis, the hour hand passes around a 12-hour clock twice. A.M. is used to identify the first 12 hours and P.M. the next 12 hours. Each hour is divided into 60 minutes. Students will be setting and reading hour and half-hour times.

Management
1. Patty paper, found in restaurant supply stores, is translucent and often comes in pre-cut squares. It is used to separate meat patties in packages.
2. Copy the clock hands and clock circle (if not using paper plates) on card stock.
3. To build the concept of angle measurement, *From Wedges to Wangles* should precede this activity.
4. When showing times on the clock, remind students the clock hands must be in the same position as would normally occur on a real clock.

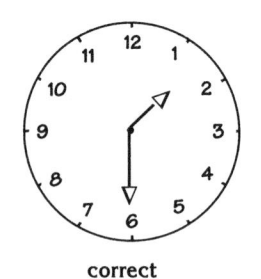

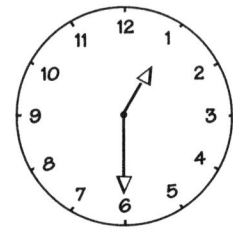

correct   incorrect

5. Choose times that are appropriate for your students—hours only or hours and half hours.

Procedure

1. Explain that the individual wedge-shaped units they have been using, nicknamed wangles, can be put edge to edge to make an angle measuring tool called a protractor. They will use the wangle protractor to explore the geometry of clocks.
2. Give students the wangle protractor page and waxed paper. Guide them through the folding process.
3. Ask, "Into how many wangles is the protractor divided?" [24] "How can we use it to make a clock?" [Since a clock is divided into 12 sections, every two wangles mark an hour.]
4. Distribute the clock hands and a paper fastener to each student. After they have determined the approximate center of the plate, have them use the wangle protractor to mark and number the clock. Attach the clock hands.
5. Instruct students to set their clocks at 3:00. Ask, "In how many ways could you measure this angle?" [2, a small measure and a large measure] Reinforce that there is only *one* angle formed by the clock hands, but it can be measured two ways. Because mathematicians define an angle as a straight line or less, they should use the smaller measure.

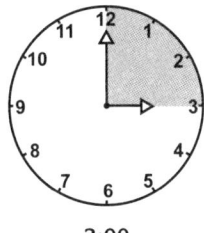

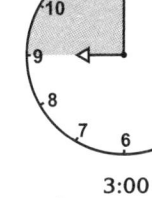

3:00 small measure   3:00 large measure

6. Give students the recording page. Name a time, say 8:00, and have them show it on their clocks. Direct students to draw the clock hands, shade the angle, and record the number of wangles on the page.
7. Ask students to find another clock angle that is congruent (same size, same shape) with the eight o'clock angle and has hand positions found on a real clock. (After some problem solving, they should discover that four o'clock is a match, a connection to reflection symmetry.) Have them record this next to the eight o'clock; notice the clocks are arranged in pairs.

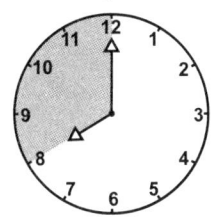

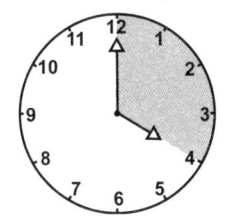

Reflection symmetry

8. Choose different students to announce a time (see *Management 5*). For each time, the class should change their clock hands, record the data, and find a matching angle.

Connecting Learning

1. What is another name for a wangle edge going out from the center of the circle? [radius]
2. What do you notice about the pairs of matching angles? [They are reflections of each other; they are symmetrical. Times reflect on either side of an imaginary line going through 12 and 6.]
3. Which clock times (hours and half hours) do not have a matching angle? [12:00 which has no angle at all and 6:00, a straight angle]
4. How many wangles wide are your scissors when completely open? Do all scissors open the same amount?
5. How does using the wangle protractor compare to using individual wangle pieces? [It's much easier to hold just one piece of paper to measure.]

Extensions

1. Use the wangle protractor to measure angles around the room. Have students list the results on a piece of chart paper. Students are likely to encounter angles that measure fractions of wangles, pointing to the need for a smaller measuring unit.
2. Have students explore the two measures of the angle formed by clock hands (see *Procedure 5*). Instruct students to try different times, record both measures in a T-table, and look for a pattern. [The small and large measures have a sum of 24 wangles, the total number of wangles in a circle.]

Time	Small Measure	Large Measure
7:30	3 wangles	21 wangles
3:00	6 wangles	18 wangles

3. Using a clock where the hour and minute hands move with each other (available from AIMS), challenge students to find other angles—beyond just hours and half-hours—that match the 6:00 angle. [There is a match within every hour.]
4. Have students investigate this question: Are there any other ways, besides wedges, to divide a circle into congruent (same size, same shape) parts?

Home Link

Have students use the wangle protractor to measure and record items at home.

(The wangle protractor was adapted from an idea by John A. Van de Walle in the book, *Elementary School Mathematics: Teaching Developmentally*. Longman. White Plains, New York. 1990.)

* Reprinted with permission from *Principles and Standards for School Mathematics*, 2000 by the National Council of Teachers of Mathematics. All rights reserved.

Waxed Wangles

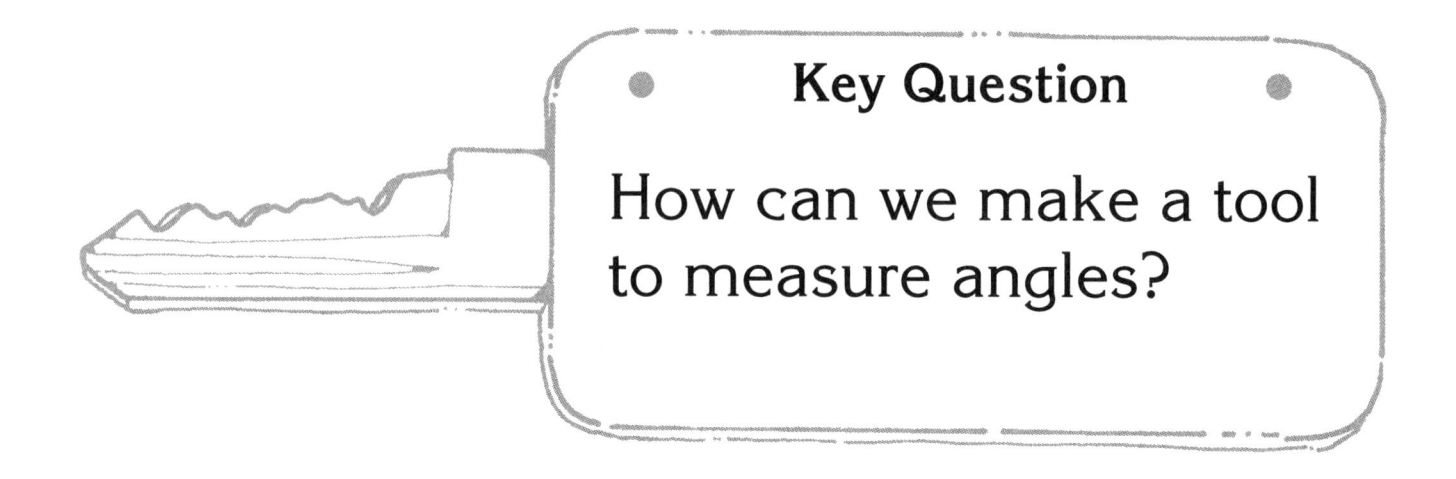

Key Question

How can we make a tool to measure angles?

Learning Goals

Students will:

- fold a non-customary protractor, and
- use it to construct a clock and measure angles formed by the clock's hands

Waxed Wangles

Wangle Protractor

- Fold a square of waxed paper in half **three** times.
- Place over the template and fold again at the dashed lines.
- Cut and open.

template

Use your wangle protractor to make a clock. Attach clock hands.

HARDHATTING IN A GEO-WORLD © 2004 AIMS Education Foundation

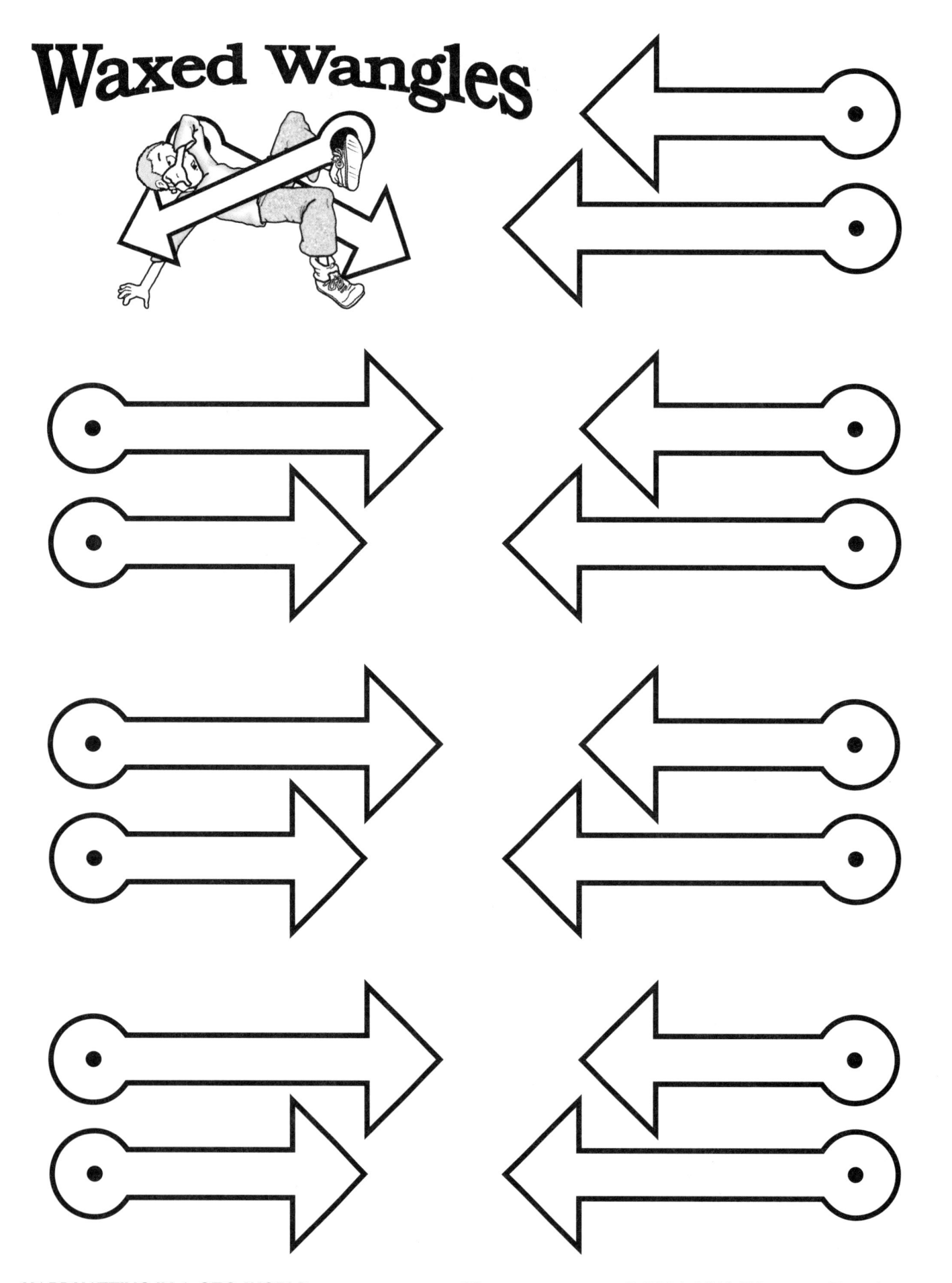

Waxed Wangles

HARDHATTING IN A GEO-WORLD 98 © 2004 AIMS Education Foundation

Waxed Wangles

Show a time on the clock.
Draw the hands, shade the angle,
and record the number of wangles.

HARDHATTING IN A GEO-WORLD 99 © 2004 AIMS Education Foundation

Waxed Wangles

1. What is another name for a wangle edge going out from the center of the circle?

2. What do you notice about the pairs of matching angles?

3. Which clock times (hours and half hours) do not have a matching angle?

4. How many wangles wide are your scissors when completely open? Do all scissors open the same amount?

5. How does using the wangle protractor compare to using individual wangle pieces?

WANGLE-ROUND UP

Topic
Building a protractor with non-customary units

Key Question
How can wangles be used to make a measuring tool?

Learning Goals
Students will
- construct a circular protractor with non-customary units, and
- use it both for measuring and drawing angles.

Guiding Documents
Project 2061 Benchmark
- *Use numerical data in describing and comparing objects and events.*

NCTM Standards 2000*
- *Understand the need for measuring with standard units and become familiar with standard units in the customary and metric systems*
- *Understand that measurements are approximations and understand how differences in units affect precision*
- *Select and apply appropriate standard units and tools to measure length, area, volume, weight, time, temperature, and the size of angles*
- *Use geometric models to solve problems in other areas of mathematics, such as number and measurement*

Math
Geometry and spatial sense
Measurement
    angle

Integrated Processes
Observing
Collecting and recording data
Comparing and contrasting

Materials
Card stock for the protractor
Transparencies for the rays
Scissors
1/8" hole punches (see *Management 2*)
Size 3/0 snaps or paper fasteners, 3 per student
    (see *Management 3*)
3-D solids (see *Management 4*)
Metric ruler
Transparency of the line design page

Background Information
The protractor is a poorly understood measuring tool. This is because the measuring unit (a degree) is very small, the measuring unit's wedge shape is not visible, and the protractor has a confusing double-numbering scale.[1]

The unit used to measure an angle is wedge-shaped. In this activity, students construct a protractor with large wedge-shaped units, nicknamed wangles, that gives them an opportunity to make sense of protractors. Students can see that the protractor is made of repeated units in the same way multiple centimeters are used to form a meter stick. Later that understanding can be transferred to a very small measurement unit called a degree.

This protractor also has visible rays to help cement the image of the wedge-shaped measuring unit in their minds. Being relatively large, wangles are not too laborious to count so a numbering scale is not needed initially. When the need arises, the *Extension* provides a procedure for introducing the double numbering system.

Both the waxed paper protractor *(Waxed Wangles)* and this wangle protractor are circular. Students should understand that the angle measuring unit, whether a wangle or degree, sweeps completely around the center point of a circle. Connections can later be made to the 360° of a circle and to the use of directional compasses. This circular concept should be developed before transitioning to degree protractors, which are most often semi-circular.

Wangles are large. The need for a smaller measure is deliberately addressed in two ways. On the *3-D Models* page, the angles of the triangular face on the square pyramid model do not measure a whole number of wangles; they are in-between. The same will likely be true of the triangle students draw on the *Quilt* page. The need for a more precise measuring unit leads directly to degrees.

[1]. Van de Walle, John A. *Elementary School Mathematics: Teaching Developmentally.* Longman. White Plains, New York. 1990.

Management
1. Each student will need one protractor circle and two transparent rays. Copy the protractor construction page on card stock and the rays page (sufficient for eight students) on transparencies.
2. Eighth-inch hole punches, found at stationery and school supply stores, are the preferred tool for making the small holes necessary for protractor assembly. They are safe and leave a clean hole, important for the unobstructed movement of one of the rays. It is helpful to have several to be shared by the class.

HARDHATTING IN A GEO-WORLD   101   © 2004 AIMS Education Foundation

3. Since the rays are transparent, snaps offer less eye clutter than paper fasteners. This is particularly important for the center hole or vertex. Snaps work best when the hole is clean; otherwise they tend to pop off. One option is to use paper fasteners for the side holes and a snap for the center hole.
4. Notice that students are not asked to number the protractor; this is deliberate. Knowing which of the two scales to read on a protractor, a confusing issue for students, follows understanding how a protractor is used to measure. With a little experience, however, they may begin to move naturally toward numbering. Grasp this teachable moment by following the suggestions in the *Extension*. It is preferable to address the double system using large units (wangles) rather than very small units (degrees).
5. Three solids are needed for this activity: a triangular prism, a hexagonal prism, and a square pyramid. Each set of models can be shared by two or three students. If you have the Power Solids (available through AIMS), use those. If not, the *Quilt* page is provided as a 2-D alternative.
6. It is helpful to have a transparency of the *Line Designs* page to illustrate how they are constructed.

Procedure
1. Explain that students will be building their own protractor to measure angles. Distribute the construction page, transparent rays, and snaps. Have students follow the construction directions, noticing the wedge-shape units used to mark the protractor.
2. Say, "Show me a 4-wangle angle." Have students count and move the protractor ray to the correct position. Explain that neighboring or adjacent angles share a ray. The measure on one side of the ray is 4 wangles. "What is the measure on the other side?" [8 wangles (Mathematicians define angles as measuring 180° or less—12 wangles or less in this case—so only the half-circle should be considered.)]

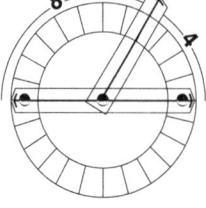

Adjacent angles

Repeat using other angle measures. Then ask, "How many wangles are in a half-circle? [12 wangles] …in a whole circle?" [24 wangles]
3. Give students the 3-D models and corresponding page. (If these models are not available, skip to *Procedure 6*.) Ask, "How many faces does a triangular prism have?" [5] "What shapes do the faces have? [triangle, square] Direct students to trace these two shapes on the page.
4. Have students place the triangular prism on the protractor, lining up the vertex of one of its faces with the center of the protractor and moving the free ray until the protractor angle matches the edges of the face.

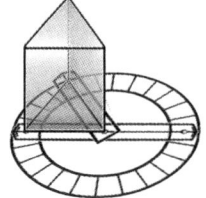

5. Instruct students to count the number of wangles in each angle and record them on the traced faces. Repeat for the hexagonal prism and the square pyramid.

As questions arise when students discover angles that do not fall on wangle lines, ask, "How should the number of wangles be recorded?" (One suggestion is to add a plus sign as in 3+, meaning more than 3 wangles but less than 4.)

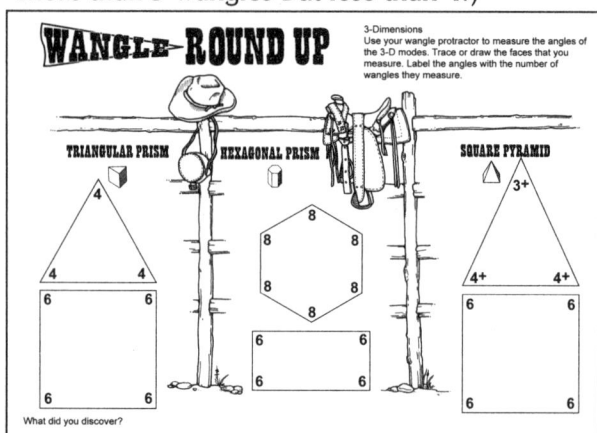

6. Distribute the *Quilt* page, particularly if students did not work with the 3-D models. Inform them that only one of each of the three different shapes needs to be measured. Have them follow the directions on the page.

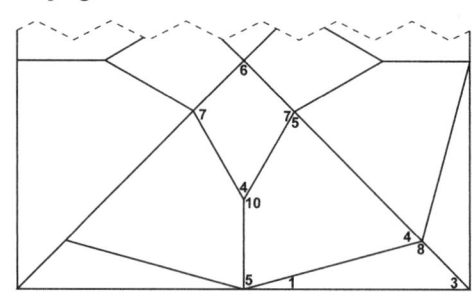

7. Give students the page for constructing angles, another protractor skill. On an overhead projector, demonstrate how to place the protractor's vertex at the left end of the acute angle line. Then line up the protractor's base ray with that same line. Make a mark at the three-wangle line and remove the protractor. Take a ruler and draw a line from the vertex through the mark.

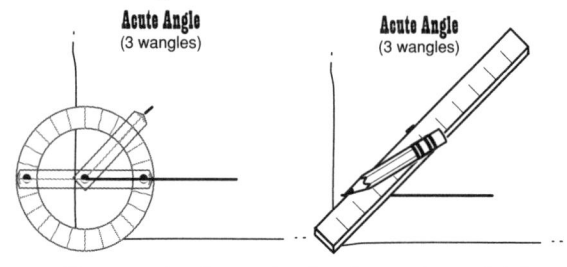

Acute Angle (3 wangles)   **Acute Angle** (3 wangles)

Use a ruler to make hash marks, which make more precise intersections than points, at every centimeter along the 7-cm base line. Do the same

HARDHATTING IN A GEO-WORLD

on the newly-drawn line. If the line is longer than 7 cm, erase the extra length. If the line is shorter, extend it to 7 cm.

Show students how to number and connect hash marks. Once they understand the pattern, they may not need to number the other angles. Share the secret of maintaining a straight line while drawing: Spread the thumb and first finger as wide as possible and press against the ruler. With pressure in two places, the ruler is less likely to shift.

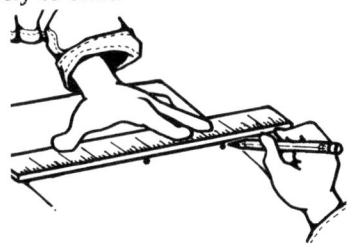

8. Have students work on the three line designs, each time constructing the angle with the vertex on the *left* side of the line. This consistent pattern aids comparison. Also direct them to write the number of wangles in a right angle.
9. Ask students to compare and discuss the results.

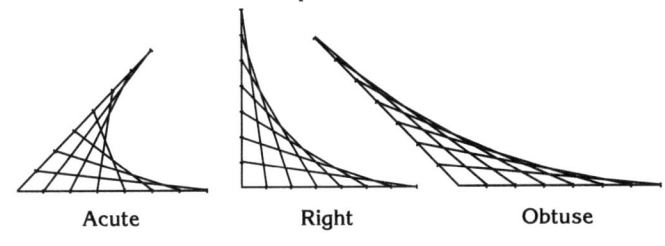

Acute   Right   Obtuse

Connecting Learning
3-D Models
1. What do you notice about the angles of the square? [They are all equal, all 6 wangles.] ... the angles of the triangle? [4 wangles] ... rectangle? [6 wangles, the same as the square] ... hexagon? [8 wangles]
2. What did you discover? [The angles of the triangle on the square pyramid can't be measured in whole wangles.] How should the number of wangles be recorded? (possibly 3+ and 4+)
3. How could the protractor be improved to solve this measuring problem? [Smaller units are needed to measure more precisely.]

Quilt
1. What shapes do you see and what are their wangle measures? [kite—6, 7, 4, 7; general quadrilateral—5, 4, 5, 10; triangle (scalene, obtuse)—8, 3, 1]
2. If you rotate the quilt in one complete circle, how many times is the pattern repeated? [4 (It has rotational symmetry.)]
3. What did you discover while measuring the triangle? [Smaller units are needed to measure more precisely.]

Line Designs
1. Since the wangle measure wasn't given, what was your thinking about how to make the right angle? How many wangles is it? [6 wangles]
2. How does the size of the angle affect the line design pattern? [The smaller the angle, the sharper the curve. The larger the angle, the more the curve is stretched out or the closer it comes to a straight line.]

Extension
Introducing numbering scales
Even though wangles are rather large, students may tire of counting. Have them suggest a solution (numbering) and then test it by measuring objects.

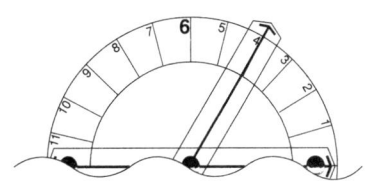

Students may soon realize the protractor would be more versatile if angles could be measured from either the left side or the right side. A second scale, starting with zero at the opposite end of the protractor, would meet this need.

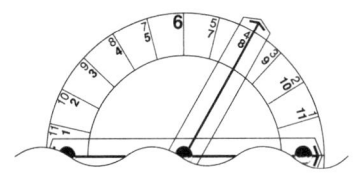

Have students practice using the double numbering scale. Which scale to use makes more sense when it arises out of a perceived need, is presented with larger units (wangles), and is written in by the students themselves.

* Reprinted with permission from *Principles and Standards for School Mathematics*, 2000 by the National Council of Teachers of Mathematics. All rights reserved.

WANGLE ROUND UP

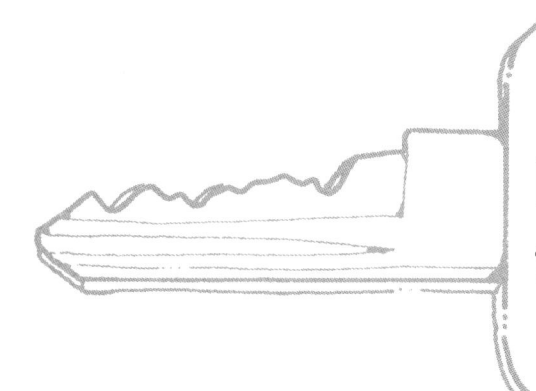

Key Question

How can wangles be used to make a measuring tool?

Learning Goals

Students will:

- construct a circular protractor with non-customary units, and
- use it both for measuring and drawing angles.

WANGLE-ROUND UP   Construction

1. Cut both the outer and inner circles of the protractor.

2. Cut the two transparent arrows just inside the black-line border. Only the center line and ray should be seen.

3. Punch one-eighth inch holes, where shown, in both the transparent arrows and the circle. Use snaps or paper fasteners to attach the transparent diameter across the circle. Add the ray to the center of the protractor.

HARDHATTING IN A GEO-WORLD   105   © 2004 AIMS Education Foundation

WANGLE ROUND UP

Make transparencies.

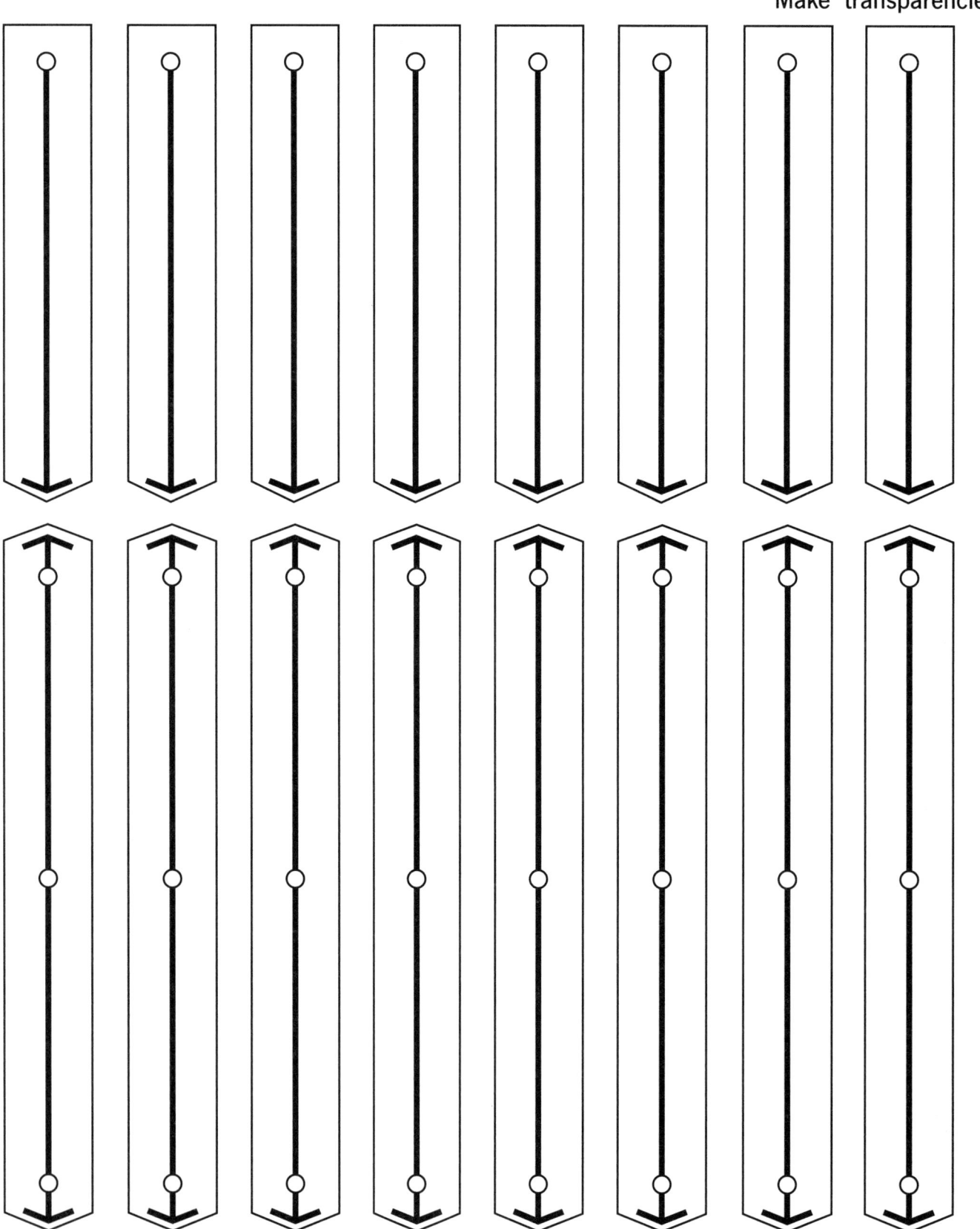

HARDHATTING IN A GEO-WORLD 106 © 2004 AIMS Education Foundation

WANGLE ROUND UP

3-D Models

Use your wangle protractor to measure the angles of the 3-D models. Trace or draw the faces that you measure. Label the angles with the number of wangles they measure.

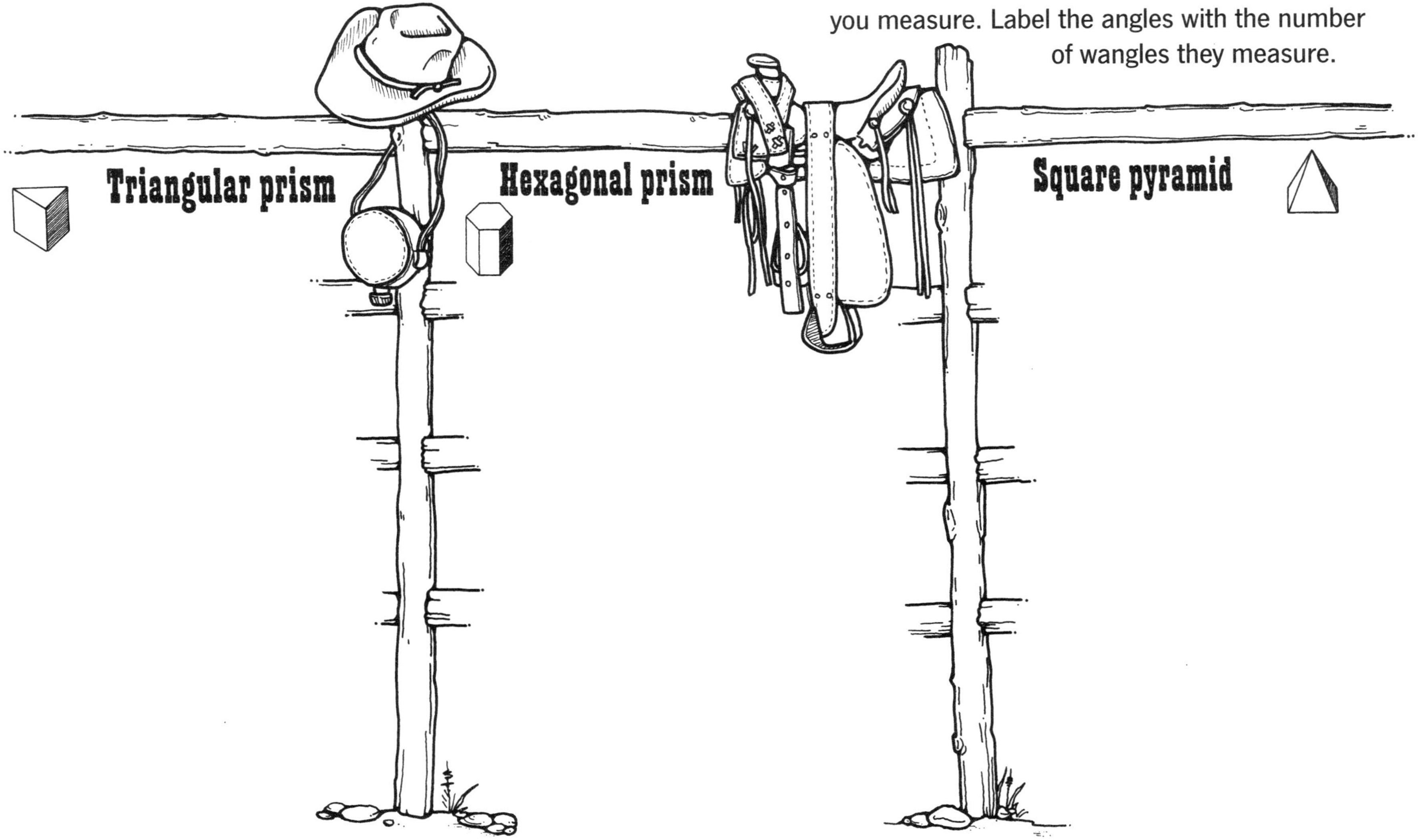

Square pyramid

Hexagonal prism

Triangular prism

What did you discover?

HARDHATTING IN A GEO-WORLD 107 © 2004 AIMS Education Foundation

WANGLE ROUND UP

Quilt

This quilt square is made of only three shapes. Use your wangle protractor to measure the angles of these three shapes. Record at the angles.

Color the shape that has a 7-wangle angle _____.

Color the shape that has a 5-wangle angle _____.

Color the shape that has an 8-wangle angle _____.

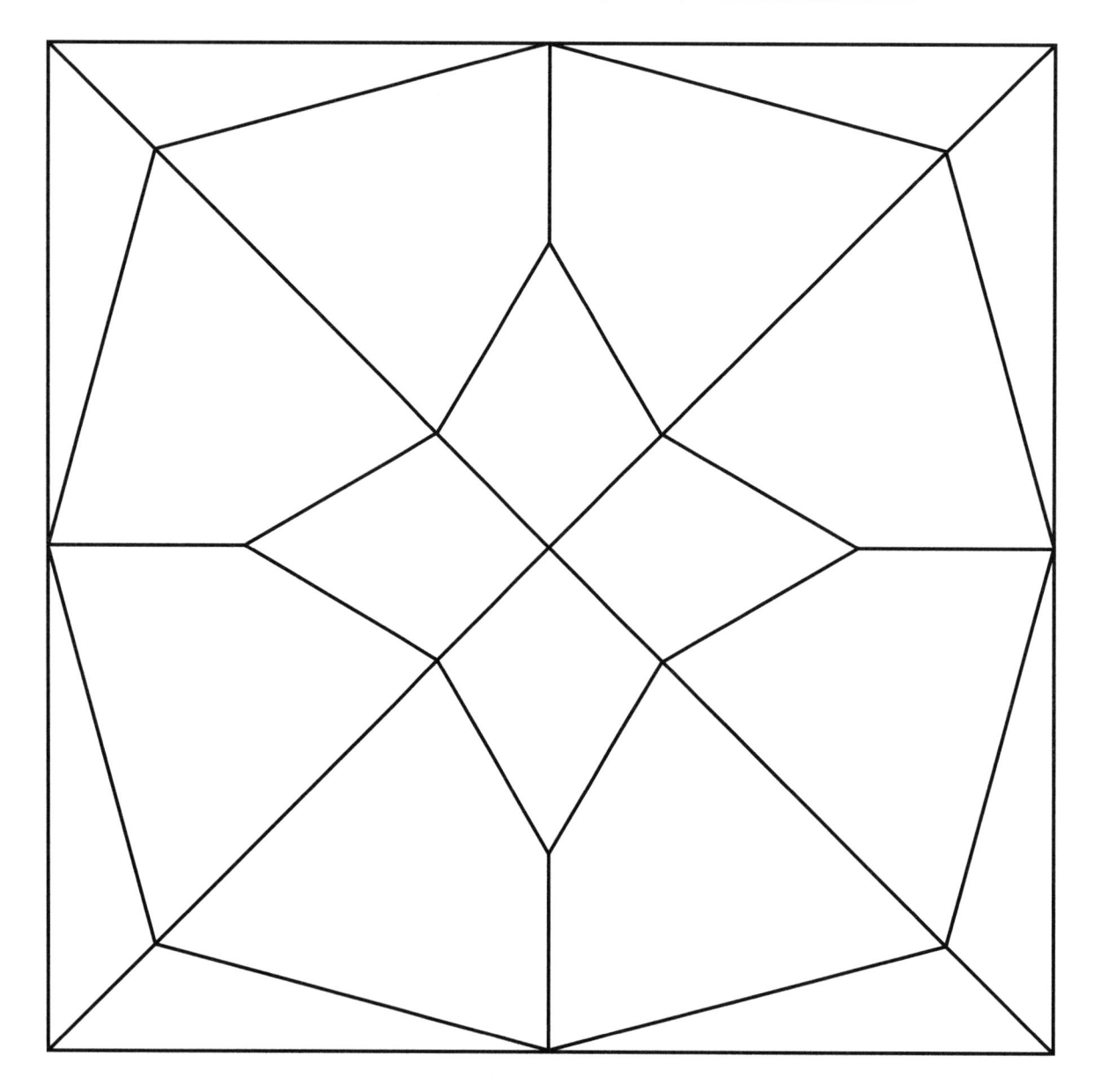

On the back of this paper, use a straight edge to draw any kind of triangle. Then measure the angles. What did you discover?

HARDHATTING IN A GEO-WORLD © 2004 AIMS Education Foundation

WANGLE ROUND UP

Line Designs

How do the line design patterns compare?

Use your wangle protractor to draw three angles, each with rays 7 cm long. Mark every centimeter along each ray and number like the illustration. To complete the line designs, use a straight edge to draw lines between matching numbers.

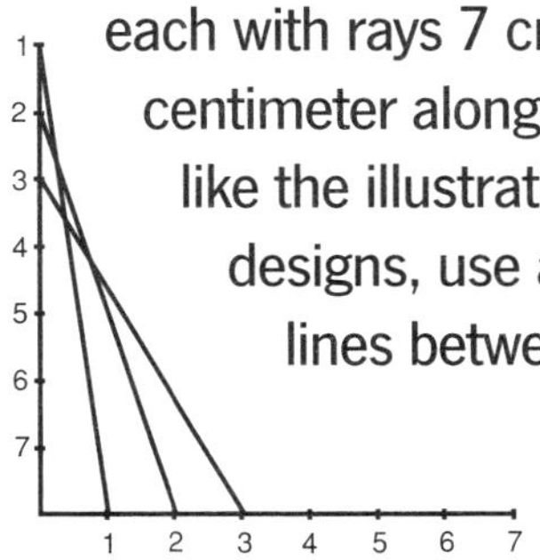

Obtuse Angle
(9 wangles)

Right Angle
(___ wangles)

Acute Angle
(3 wangles)

HARDHATTING IN A GEO-WORLD   109   © 2004 AIMS Education Foundation

WANGLE ROUND UP

3-D Models
1. What do you notice about the angles of the square? ...the angles of the triangle? ...rectangle? ...hexagon?
2. What did you discover? How should the number of wangles be recorded?
3. How could the protractor be improved to solve this measuring problem?

Quilt
1. What shapes do you see and what are their wangle measures?
2. If you rotate the quilt in one complete circle, how many times is the pattern repeated?
3. What did you discover while measuring the triangle?

Line Designs
1. Since the wangle measure wasn't given, what was your thinking about how to make the right angle? How many wangles is it?
2. How does the size of the angle affect the line design pattern?

Shaping Up

Topic
Geometric shapes

Key Question
How many different geometric shapes can you find as we walk around the neighborhood?

Learning Goal
Students will find and draw examples of geometric attributes in the real world.

Guiding Documents
Project 2061 Benchmarks
- *Numbers and shapes—and operations on them—help to describe and predict things about the world around us.*
- *Shapes such as circles, squares, and triangles can be used to describe many things that can be seen.*
- *Many objects can be described in terms of simple plane figures and solids. Shapes can be compared in terms of concepts such as parallel and perpendicular, congruence and similarity, and symmetry. Symmetry can be found by reflection, turns, or slides.*

NRC Standard
- *Objects have many observable properties, including size, weight, shape, color, temperature, and the ability to react with other substances. Those properties can be measured using tools, such as rulers, balances, and thermometers.*

NCTM Standards 2000*
- *Build and draw geometric objects*
- *Classify two- and three-dimensional shapes according to their properties and develop definitions of classes of shapes such as triangles and pyramids*

Math
Geometry and spatial sense
  dimensions (1-D, 2-D, 3-D)

Integrated Processes
Observing
Classifying
Collecting and recording data
Comparing and contrasting

Materials
Board or book to support paper while drawing
Crayons or colored pencils

Background Information
Geometric shapes are everywhere, from the foods we eat to the buildings in which we live and work. Shapes surround us at school, in the neighborhood, on the horizon, and in space.

We can gain an awareness and appreciation of the geometry in our world through observation. The more focused our observations become, the more details we notice. We then become able to connect the language of geometry to all sorts or real-world settings.

To aid students in their exploration, some geometric attributes, grouped by dimension, are listed on the first activity page. A line itself has one dimension, length, though it can meander through a two-dimensional plane or three-dimensional space. Two-dimensional shapes have length and width. Three-dimensional objects have length, width, and height.

Management
1. A walk in the surrounding neighborhood is suggested, but it could also be done around the school grounds.
2. Identify the attributes for which you want students to search. For instance, you may want to concentrate only on examples of *parallel* and *perpendicular*. At another time, the page can be used for a different search.
3. The first activity page deals with finding examples of particular geometric attributes. The second page focuses more on objects and identifying the multiple attributes they have.

HARDHATTING IN A GEO-WORLD 111 © 2004 AIMS Education Foundation

Procedure

1. Explain that the class will be taking a geometry walk. They will be looking for lines and shapes found in nature and in objects made by people.
2. Give students the first activity page and a firm writing surface such as a clipboard or book. Have them underline or circle the attributes on which you want them to focus.
3. Guide the walk, stopping frequently to give students time to draw and label their examples.
4. Return to the classroom and have students share what they found.
5. Now or on another day, distribute the second activity page.
6. Take students outside and challenge them to find objects that have multiple geometric attributes. Have them write the object's name and attributes (from the first page, if needed) in the spaces provided. Suggest they draw a picture too.

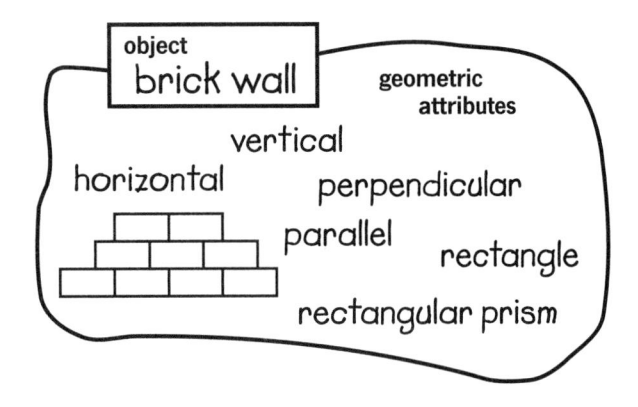

[Diagram: object "brick wall" with geometric attributes: vertical, horizontal, perpendicular, parallel, rectangle, rectangular prism]

7. Invite students to share the objects they chose and the attributes they observed.

Connecting Learning

1. What attributes were hardest to find? …easiest to find?
2. How many different objects can we name that have squares? …cubes? …horizontal lines?, etc. (Make lists on chart paper or a transparency.)
3. In how many different ways can we describe a toy wagon using geometric words?
4. What object did you find most pleasing to observe? What attributes does it have?

Curriculum Correlation

Language Arts
1. Hoban, Tana. *Spirals, Curves, Fanshapes and Lines*. Greenwillow Books. New York. 1992. (This book has colorful photographs of the geometry in our world.)
2. Have students write poetry in geometric shapes.
   a. Write a poem with three lines (one word, two words, three words) in the shape of a triangle or with four lines (four words per line) in the shape of a square.

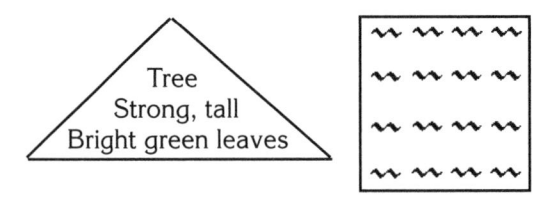

[Triangle poem: Tree / Strong, tall / Bright green leaves]

   b. Write a poem about squares inside a square, a poem about triangles inside a triangle, etc.

Art
1. Make a picture using only geometric shapes.
2. Create a collage using one shape of various sizes and colors, or use a combination of one line and one shape.
3. Build three-dimensional shapes out of construction paper.

Home Link

Have students do a shape hunt around their house.

* Reprinted with permission from *Principles and Standards for School Mathematics*, 2000 by the National Council of Teachers of Mathematics. All rights reserved.

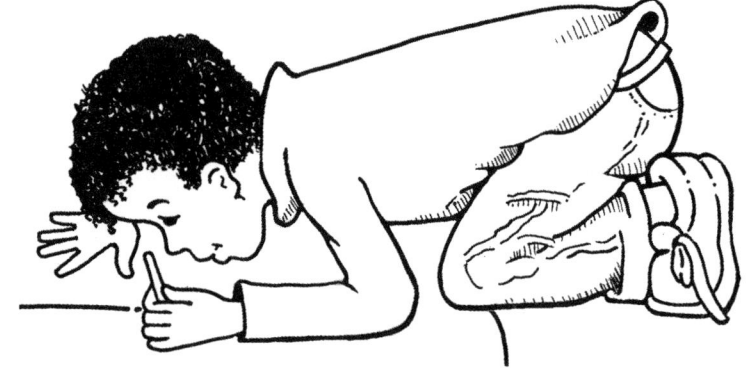

HARDHATTING IN A GEO-WORLD

Shaping Up

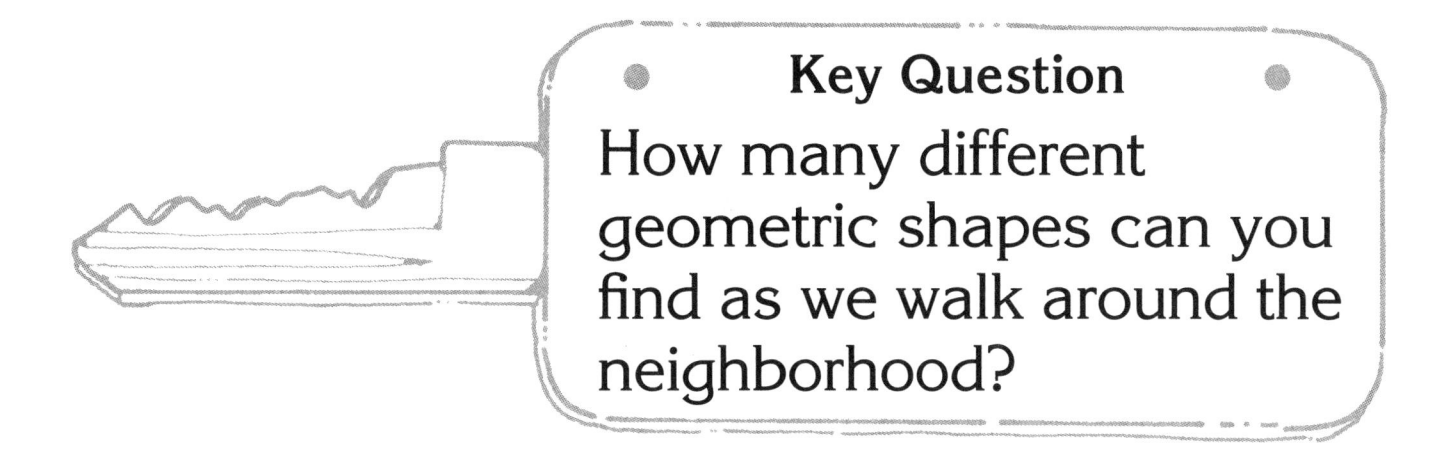

Key Question

How many different geometric shapes can you find as we walk around the neighborhood?

Learning Goal

Students will:

- find and draw examples of geometric attributes in the real world.

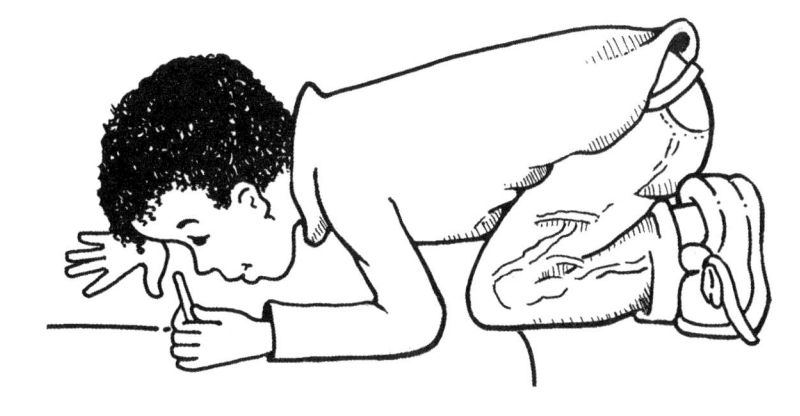

Shaping Up

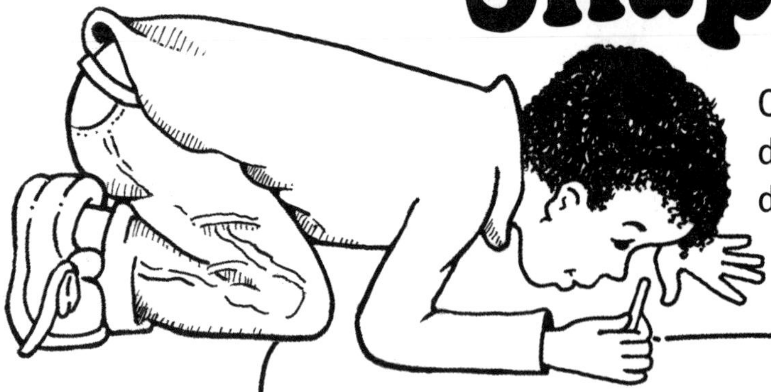

Circle the geometric attributes you will look for during your walk. Draw and label natural and designed objects with these attributes.

Lines
horizontal
vertical
diagonal
parallel
perpendicular
oblique
curve

2-D Shapes
triangle
square
rectangle
pentagon
hexagon
octagon
circle
ellipse

3-D Shapes
cube
rectangular prism
tetrahedron
square pyramid
cone cylinder
sphere
hemisphere

HARDHATTING IN A GEO-WORLD © 2004 AIMS Education Foundation

Shaping Up

Find four natural and/or designed objects.
List several geometric attributes for each one.

object

geometric attributes

object

geometric attributes

object

geometric attributes

object

geometric attributes

HARDHATTING IN A GEO-WORLD   115   © 2004 AIMS Education Foundation

Shaping Up

CONNECTING LEARNING

1. What attributes were hardest to find? ...easiest to find?

2. How many different objects can we name that have squares? ...cubes? ...horizontal lines?, etc.

3. In how many different ways can we describe a toy wagon using geometric words?

4. What object did you find most pleasing to observe? What attributes does it have?

Slice Me Twice

Topic
Quadrilaterals

Key Question
What happens when we cut circles that are attached to each other?

Learning Goal
Students will investigate how 3-D circle constructions are changed into 2-D quadrilaterals.

Guiding Documents
Project 2061 Benchmarks
- *Mathematics is the study of many kinds of patterns, including numbers and shapes and operations on them. Sometimes patterns are studied because they help to explain how the world works or how to solve practical problems, sometimes because they are interesting in themselves.*
- *Many objects can be described in terms of simple plane figures and solids. Shapes can be compared in terms of concepts such as parallel and perpendicular, congruence and similarity, and symmetry. Symmetry can be found by reflection, turns, or slides.*

NRC Standards
- *Employ simple equipment and tools to gather data and extend the senses.*
- *Communicate investigations and explanations.*

NCTM Standards 2000*
- *Create and describe mental images of objects, patterns, and paths*
- *Investigate, describe, and reason about the results of subdividing, combining, and transforming shapes*
- *Identify, compare, and analyze attributes of two- and three-dimensional shapes and develop vocabulary to describe the attributes*

Math
Geometry
Spatial visualization

Integrated Processes
Predicting
Observing
Collecting and recording data
Comparing and contrasting
Generalizing
Relating

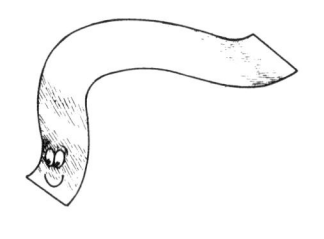

Materials
Paper strips (see *Management 1*)
Transparent tape
Scissors

Background Information
The power of this activity lies in exploration and discovery. Students are introduced to some specific two-circle constructions to investigate and then are encouraged to test ideas of their own. All of the constructions move from three-dimensions (length, width, height) to two-dimensions (length and width).

Spatial visualization is an important component of the prediction process. Students need to picture in their minds how the way the circles are constructed will determine the shape that results when the circles are cut. As more data are collected, a more accurate prediction should be possible.

Properties of Two-Circle Shapes

Rectangle	Square
four sides	four sides
opposite sides equal	all sides equal
opposite sides parallel	opposite sides parallel
four right angles	four right angles

Parallelogram	Rhombus
four sides	four sides
opposite sides equal	all sides equal
opposite sides parallel	opposite sides parallel
opposite angles equal	opposite angles equal

Management
1. Cut paper strips, at least 1" x 11", in two colors. Plan on eight or more strips for each student, depending on how many explorations will be done.
2. Taping along entire edges and on both sides where strips attach to each other is extremely important.

(The following is offered for those students ready for more independent work. With either approach, have students write a description/result of one of their constructions. Mount these on a wall and challenge students to find ones not yet discovered.)

Open-ended: Challenge students to explore the *Key Question*, organize their results, and offer conclusions relating the way the circles are constructed to the shape produced when they are cut.

Guided planning: Give students the *Plan Sheet* to help them organize their investigation of the *Key Question*.

HARDHATTING IN A GEO-WORLD © 2004 AIMS Education Foundation

Procedure

1. Give each student two equal paper strips, one of each color. Transparent tape and scissors should be available.
2. Direct students to loop each strip into a circle and secure by taping completely across both sides.
3. Show students how to tape one circle on top of the other at 90° angles. Reach inside the top circle and tape both edges to the bottom circle. Turn it over and tape the inside edges.

4. Ask the *Key Question*. Distribute the activity sheet and have students fill in the first three columns. Have them predict the shape they will get when the circle is cut along the middle.

# of Circles	Equal or Unequal Circles	90° or 45° Angles
2	equal	90°

5. Show students how to slightly pinch one circle, snip the center, and cut all the way around the middle of the circle.
6. Instruct students to cut one of their circles. They will have a straight strip with a loop at each end. They should cut down the center of the straight piece and lay the figure flat.

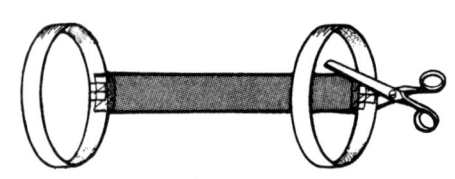

7. Once the shape is recorded, discuss the properties of the square together.
8. Ask, "What will happen if we start with one circle larger than the other?"
9. Distribute more strips. Have students shorten one strip and repeat the process.
10. Ask, "What will happen if we tape the circles at a different angle?" Show them how to tape circles at a 45° angle. Then let them explore the shapes made with equal and unequal circles.
11. Discuss what has been discovered.
12. Ask, "What other circle constructions would you like to explore?" Give them more strips for their investigations.

Connecting Learning

1. How does the way the circles are constructed affect the shape when they are cut? [taping at 90° made shapes with right angles, same length made shapes with equal sides, etc.]
2. How are the shapes (quadrilaterals) alike?...different?
3. What other circle constructions would you like to explore?

Extensions

1. Explore the lines of symmetry and/or the characteristics of the diagonals for each quadrilateral.
2. Try other constructions such as a circle taped within a circle, three circles, four circles, etc. A particularly intriguing challenge would be to make a circle construction that will produce a triangle(s). [Two equilateral triangles (with twists) can be made by equally spacing three circles within each other. Attach only on one end of the sphere-like shape.]

Three circles

Construction for triangles

3. Make circle constructions using paper with differently colored sides (origami paper, wrapping paper, etc.). Compare the color placement of the construction to the results after cutting. Encourage students to find a way to make a square (or other shape) with a frame showing the same color or pattern on one side.
4. Challenge students to construct a picture frame, with paper strips, that will fit one of their pieces of art work.

Curriculum Correlation

Language Arts

Write *Who Am I?* riddles using the properties of the quadrilaterals. Example: "I have four equal sides and four right angles. Who am I?"

* Reprinted with permission from *Principles and Standards for School Mathematics*, 2000 by the National Council of Teachers of Mathematics. All rights reserved.

Slice Me Twice

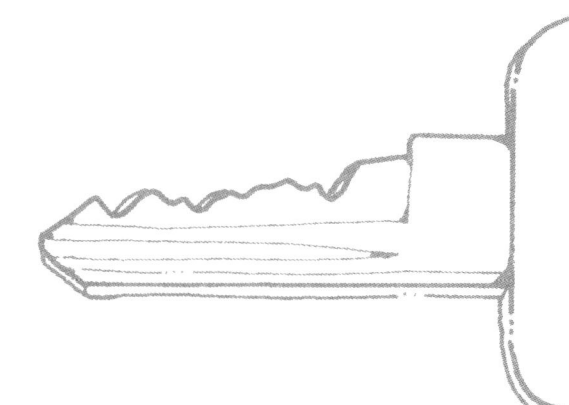

Key Question

What happens when we cut circles that are attached to each other?

Learning Goal

Students will:

- investigate how 3-D circle constructions are changed into 2-D quadrilaterals.

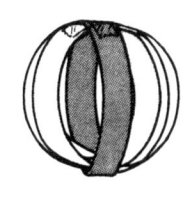

Slice Me Twice

Make circles from paper strips. Tape them together at 90° or 45° angles. Fill in the first three columns. Predict the shape that will result when the circles are cut along the middle. Try it!

# of circles	Equal or unequal circles	90° or 45° angles	Predicted shape	Actual shape	Properties of shape

HARDHATTING IN A GEO-WORLD

Slice Me Twice

1. How does the way the circles are constructed affect the shape when they are cut?

2. How are the shapes (quadrilaterals) alike? ...different?

3. What other circle constructions would you like to explore?

Möbius Bands

Topic
Möbius bands

Key Question
How do combinations of twists and cuts affect the Möbius bands?

Learning Goal
Students will explore the Möbius band by observing the results of varying the number of twists and kind of cuts.

Guiding Documents
Project 2061 Benchmark
- Mathematics is the study of many kinds of patterns, including numbers and shapes and operations on them. Sometimes patterns are studied because they help to explain how the world works or how to solve practical problems, sometimes because they are interesting in themselves.

NRC Standard
- Employ simple equipment and tools to gather data and extend the senses.

*NCTM Standards 2000**
- Create and describe mental images of objects, patterns, and paths
- Investigate, describe, and reason about the results of subdividing, combining, and transforming shapes

Math
Geometry and spatial sense
  topology
Spatial visualization
Fractions

Integrated Processes
Predicting
Observing
Collecting and recording data
Comparing and contrasting

Materials
Paper strips (see *Management 1*)
Transparent tape
Scissors
Crayons or colored pencils

Background Information
A Möbius strip or band is a one-sided surface formed by giving a rectangular strip a half twist before joining the two ends together. It was discovered in 1858 by August Ferdinand Möbius (1790-1868), a German mathematician and astronomer. To illustrate its one-sidedness, a line can be drawn from any point on the band to any other point without crossing an edge.

An odd number of half twists will make a one-sided figure. An even number of half twists gives a two-sided figure which is not, by definition, a Möbius band. However, it is interesting to explore the results of cutting both odd and even numbers of half twists and generalize about the number of half twists needed to make a Möbius band.

While, at first glance, the Möbius band seems to be just another interesting mathematical oddity, it has applications in technology. During the time that Möbius discovered the band, the Industrial Revolution was in full swing in Europe and the United States. Many of the factories had a single power source, usually a steam engine or water wheel, that turned a long shaft, or series of shafts. The individual pieces of equipment in the factory were connected to the shafts by a series of belts and wheels. In this way a single steam engine in a cloth-weaving factory, for example, could be used to power dozens of looms. The belts connecting the individual machines to the overhead shaft were constantly turning and would have to be replaced periodically.

Plant engineers found that putting a half twist in the belts—making them into Möbius bands—made them last longer, since the belts would have to go around twice to get back to the same point of wear. Thus, Möbius' discovery was quickly applied to the technology of the day. Möbius bands are still used in factories today.

Management
1. Möbius bands can be made with any reasonable size or type of paper. Adding machine tape 18-24 inches long is easy for students to handle. Cut at least five strips for each student.
2. When forming the bands, always tape completely across both sides.
3. To make comparisons more easily, have students label the bands (after cutting) with the number of half twists and kind of cut.

HARDHATTING IN A GEO-WORLD © 2004 AIMS Education Foundation

(The following is offered for those students ready for more independent work.)

> *Open-ended*: Introduce a Möbius band. Encourage students to ask "What if…?" questions that can be tested. Have student groups plan how they will record and report their findings.
>
> *Guided planning*: After introducing as in *Open-ended*, guide groups with the following questions.
> - What variables will you test?
> - What kind of cuts do you want to try?
> - How will you test for one-sidedness?
> - How will you record your data, including predictions?
> - What are each group member's responsibilities?
> - What patterns did you discover?
> - In your opinion, which band and cut gave the most fascinating result?

Procedure
1. Distribute the activity sheet, paper strips, transparent tape, scissors, and crayons.
2. Have students label each end of the strips with letters.

A B C	A B C

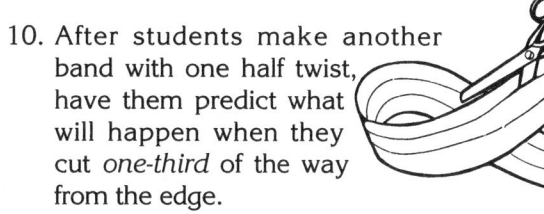

3. Direct students to use the first strip to make a band with no twists. The letters should meet on the same side. Tape the ends together completely across both sides of the strip.
4. Instruct students to take one crayon and draw down the center of the band until they connect to their starting point. Then have students take another color and do the same on the other side. Ask, "How many sides does this band have?" [Two.] Have them put this data in the table.
5. Have students predict what will happen when they cut along the line.
6. Show students how to slightly pinch the band, snip the center, and cut down the middle. Have them cut their own bands and record the results by describing or drawing.
7. Have students make a second band. Before taping, give one end a half twist. A blank side will meet a lettered side.
8. Challenge students to show the number of sides by again drawing lines with their crayons. Ask, "What did you discover?" Explain that this is a Möbius band; it has only one side.
9. Instruct students to predict what will happen when they cut down the center of the band, then make the cut and record the results. 
10. After students make another band with one half twist, have them predict what will happen when they cut *one-third* of the way from the edge.
11. Direct students to make the cut and record the results. (They will need to cut around the band two complete times.)
12. Invite students to try different numbers of half twists and cuts on their own.
13. Hold a concluding discussion.

Connecting Learning
1. Look at your bands with one half twist. How do the results of the 1/2 cut compare with those of the 1/3 cut? [1/2 cut: one band twice as long as the original, 1/3 cut: two bands, one the same length as the original and the other twice as long]
2. What patterns did you discover?
3. How are the number of half twists related to the number of sides of the band? When were the bands one-sided? [odd number of half twists makes one-sided bands, even number makes two-sided bands]
4. Which band and cut, in your opinion, gave the most fascinating result?
5. What else would you like to try?

Extensions
1. Cut a half twist band 1/4 of the way from the edge.
2. Do *Slice Me Twice* constructions with Möbius bands.
3. Teacher demonstration: Cut two identical paper strips. Place them on top of each other, make a half twist, and tape the top end to the top end and the bottom end to the bottom end. Slide a pen between the bands to show that they are two separate pieces. Then open it up. Surprised?

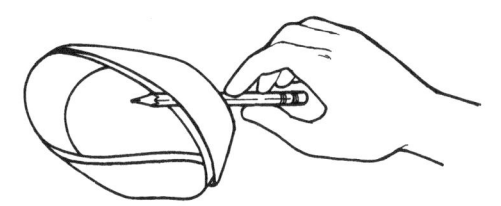

Curriculum Correlation
Literacy
Share this limerick:
   A mathematician confided
   That a Möbius band is one-sided,
   And you'll get quite a laugh
   If you cut one in half,
   For it stays in one piece when divided.
   – Author Unknown

* Reprinted with permission from *Principles and Standards for School Mathematics,* 2000 by the National Council of Teachers of Mathematics. All rights reserved.

Möbius Bands

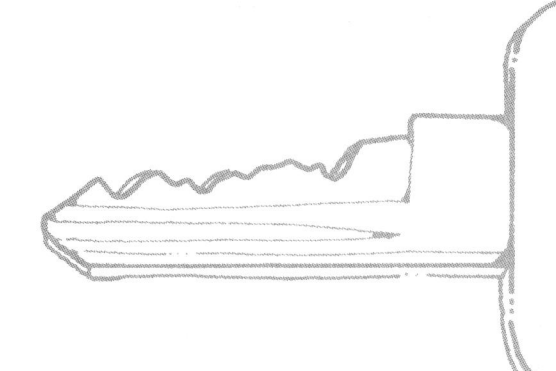

Key Question

How do combinations of twists and cuts affect the Möbius bands?

Learning Goal

Students will:

- explore the Möbius band by observing the results of varying the number of twists and kind of cuts.

Möbius Bands

How do combinations of twists and cuts affect the Möbius bands?

Start with these explorations.
Then try more of your own.

$\frac{1}{2}$ cut --------- $\frac{1}{3}$ cut =========

# of half twists	# of sides	Kind of cut	Prediction	Results (length, width, # loops)
0		$\frac{1}{2}$		
1		$\frac{1}{2}$		
1		$\frac{1}{3}$		

HARDHATTING IN A GEO-WORLD © 2004 AIMS Education Foundation

Möbius Bands

1. Look at your bands with one half twist. How do the results of the 1/2 cut compare with those of the 1/3 cut?

2. What patterns did you discover?

3. How are the number of half twists related to the number of sides of the band? When were the bands one-sided?

4. Which band and cut, in your opinion, gave the most fascinating result?

5. What else would you like to try?

GEO-PANES

Topics
3-D geometry
Minimum surfaces

Key Question
What kind of geo-pane (soap film pattern) will form on a three-dimensional shape?

Learning Goals
Students will:
- compare the properties of polyhedrons;
- look for a pattern relating vertices, face, and edges; and
- discover the soap film patterns that form on polyhedrons.

Guiding Documents
Project 2061 Benchmarks
- *Mathematics is the study of many kinds of patterns, including numbers and shapes and operations on them. Sometimes patterns are studied because they help to explain how the world works or how to solve practical problems, sometimes because they are interesting in themselves.*
- *Use numerical data in describing and comparing objects and events.*

NRC Standard
- *Think critically and logically to make the relationships between evidence and explanations.*

NCTM Standards 2000*
- *Build and draw geometric objects*
- *Identify, compare, and analyze attributes of two- and three-dimensional shapes and develop vocabulary to describe the attributes*

Math
Geometry and spatial sense
    3-D (polyhedrons)
Algebraic thinking
Patterns and relationships

Science
Physical science
    matter

Integrated Processes
Predicting
Observing
Collecting and recording data
Comparing and contrasting
Generalizing

Materials
For the class:
    liquid dish soap
    1 spool of thread or paper clips
    vinegar
    1 or more old towels, *optional*
    liter and 15 mL measures, *optional*
    newspaper and scratch paper, *optional*

For each group:
    clay (see *Management 2*)
    toothpicks
    1 container such as a half-gallon milk carton

Background Information
Geo-panes are defined as the pattern of *panes*, or soap film, created by dipping shapes into a water and soap solution. Patterns that might be expected to form around the sides of the polyhedrons meet, instead, near the center. The elastic, rubbery skin (surface tension) of the soap film stretches to cover the smallest possible area or minimum surface. Less area is covered when the soap film comes toward the center than if it were to cover the faces around the geometric shape.

The activity begins with the triangle and the square. They have two dimensions, length and width. A third dimension, height, is added when the tetrahedron, cube, triangular prism, and pyramid are built. These three-dimensional shapes are also known as polyhedrons, many-sided figures.

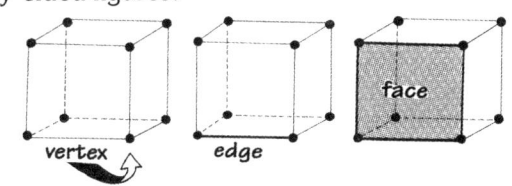

Students might also explore the relationship between vertices, edges, and faces for the polyhedrons they made. The formula, called Euler's (pronounced oilers) Theorem, states that **vertices + faces = edges + 2**. This is true for all convex polyhedrons. These are polyhedrons whose frameworks, if covered, have no indentations.

	2-D		3-D			
Shapes	△	▢	△	▢	△	△
# of Vertices (points)	3	4	4	8	6	5
# of Edges (line segments)	3	4	6	12	9	8
# of Faces (sides)	1	1	4	6	5	5

HARDHATTING IN A GEO-WORLD   © 2004 AIMS Education Foundation

Management
1. This activity will take about 60 to 90 minutes.
2. All figures are made using whole toothpicks.
3. To anchor the toothpicks at the vertices, use oil-based (plasticine) clay rolled into 1 cm balls. Scratch paper helps protect desks while rolling clay. You may prefer to use raisins or dry legumes (soak several hours to soften) instead of clay.
4. Groups of three or four should build the four polyhedrons shown on the activity sheet, each member being responsible for at least one.
5. Spread newspaper over the dipping area (tables or floor), preferably away from desks. Dipping can be done outside if the air is still.
6. For each container, pour water to a depth of no less than 9 cm, add a good squirt of liquid soap, and stir gently so bubbles do not form. If you prefer, have students measure about one liter of water and 15 mL of soap for each container. Another 15 mL of granulated sugar or glycerin (found in drugstores) may be added to strengthen the solution.
7. Caution students to dip carefully so the surface of the soapy water stays relatively free of foam and bubbles. Skim them off if necessary. The bubbles can make it difficult to see the pattern or can actually change it.
8. Students will want to experiment with more complex designs of their own. Suggest they use smaller sections of toothpicks for these.
9. To cut through the soapy film when cleaning up, sprinkle some vinegar on the wet areas and rub dry.

Procedure
1. Give each student a small lump of clay, about 12 toothpicks, about 50 cm of thread, and the activity sheet.
2. Have each student build a square or triangle with the toothpicks and clay. They should record the number of vertices and edges.
3. Borrow a triangle and dip it in the soapy water. Explain that, for this activity, the resulting pane will be called a *geo-pane*. Students should record that it has one face or flat surface. Repeat with the square. Students might notice that both two-dimensional shapes have one face.
4. Ask students to predict what will happen when they dip a three-dimensional shape in the soapy water. Give students a chance to verbalize their ideas before writing their predictions.
5. Instruct students to use their triangle or square to build one of the three-dimensional shapes. Each group should decide who will build which shape so that all of the polyhedrons shown are represented.
6. Have students record the number of vertices, edges, and faces for these shapes *before* they go anywhere near the soapy water. Help them determine the number of faces, if needed, as this may be a new term to them.
7. Show students how to slip the thread under one, and only one, toothpick and hold the thread by both ends as shown on the activity sheet. Do not tie the thread to the toothpick.
8. Have students carry their shapes to the dipping area and take turns completely submerging them in the soapy water. Suggest they dip each shape several times to see if the pattern stays the same. Students may want to build other shapes to try.
9. Hold a concluding discussion and have students write about what happened.

Connecting Learning
1. Which of the three-dimensional shapes started with a triangular base? [tetrahedron, maybe the triangular prism (Don't insist students use these terms.)] Which started with a square base? [cube, pyramid]
2. What other shapes did you build and test?
3. Is the geo-pane pattern the same each time? [usually, but a change in conditions—soap bubbles, air currents, etc.—can cause differences]
4. How could you change the geo-pane pattern? [change the thread position, pop one pane, blow on it, etc.]
5. How did you feel when you saw the geo-panes? Which was your favorite geo-pane pattern?
6. What other shapes would you like to try?

Extension
Challenge students to study the three-dimensional section of the table to find a relationship between vertices, edges, and faces (see *Background Information*).

Curriculum Correlation
Art
Have students construct polyhedrons such as cubes, tetrahedrons, and pyramids using paper patterns and glue. This is also a good preliminary activity.

Technology
Students might create three-dimensional geometric figures on the computer using *Logo*.

* Reprinted with permission from *Principles and Standards for School Mathematics*, 2000 by the National Council of Teachers of Mathematics. All rights reserved.

HARDHATTING IN A GEO-WORLD

GEO-PANES

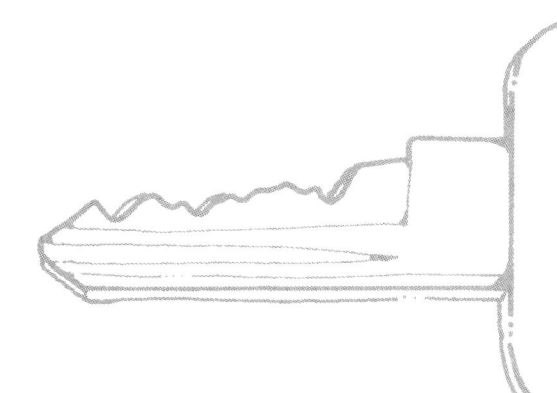

Key Question

What kind of geo-pane (soap film pattern) will form on a three-dimensional shape?

Learning Goals

Students will:

- compare the properties of polyhedrons;
- look for a pattern relating vertices, face, and edges; and
- discover the soap film patterns that form on polyhedrons.

GEO-PANES

What kind of geo-pane will form on a 3-D shape?

Build these shapes and complete the table.

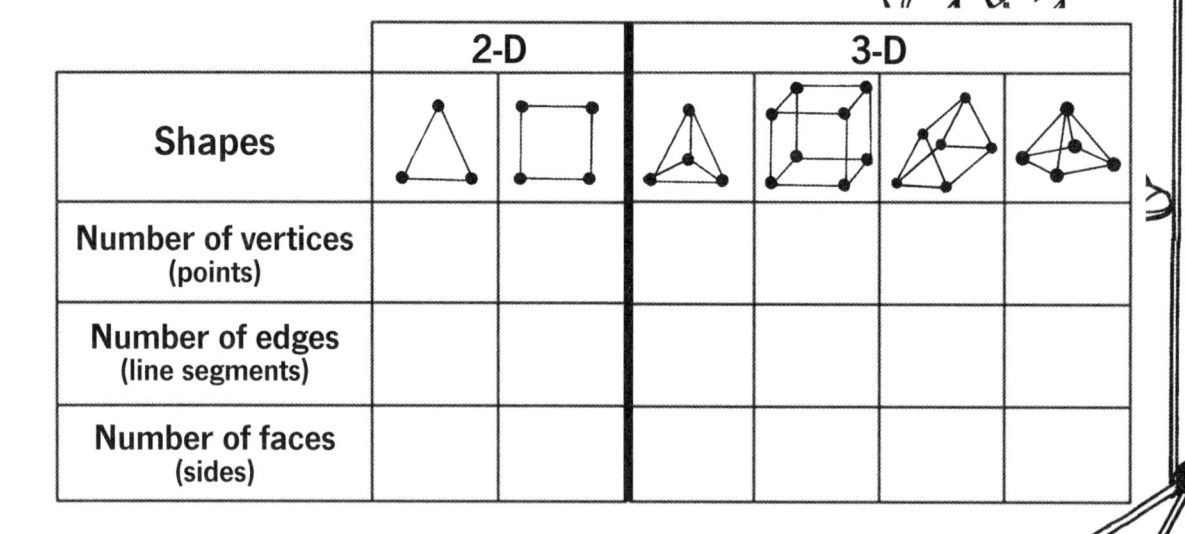

	2-D		3-D			
Shapes	△	□	△	▭	△	△
Number of vertices (points)						
Number of edges (line segments)						
Number of faces (sides)						

Hang each shape from a thread and dip completely into soapy water. Lift it out and observe.

- What do you think will happen?

- What did happen?

HARDHATTING IN A GEO-WORLD

GEO-PANES

1. Which of the three-dimensional shapes started with a triangular base? Which started with a square base?

2. What other shapes did you build and test?

3. Is the geo-pane pattern the same each time?

4. How could you change the geopane pattern?

5. How did you feel when you saw the geo-panes? Which was your favorite geo-pane pattern?

6. What other shapes would you like to try?

Edge to Edge

Topic
Polyominoes

Key Questions
1. In how many ways can squares be joined together?
2. How many different patterns can be made with five squares?

Learning Goal
Students will find as many ways as possible to combine one, two, three, four, and five squares, edge to edge.

Guiding Documents
Project 2061 Benchmarks
- Mathematics is the study of many kinds of patterns, including numbers and shapes and operations on them. Sometimes patterns are studied because they help to explain how the world works or how to solve practical problems, sometimes because they are interesting in themselves.
- Some features of things may stay the same even when other features change. Some patterns look the same when they are shifted over, or turned, or reflected, or seen from different directions.

NRC Standard
- Communicate investigations and explanations.

NCTM Standards 2000*
- Investigate, describe, and reason about the results of subdividing, combining, and transforming shapes
- Solve problems that arise in mathematics and in other contexts

Math
Geometry and spatial sense
    2-D
Problem solving

Integrated Processes
Observing
Predicting
Collecting and recording data
Comparing and contrasting

Materials
Scissors
Crayons or colored pencils
Square tiles, optional (see *Management 1*)
Envelopes, optional (see *Management 2*)

Background Information
In how many ways can squares be joined together? Following certain rules (see below), this question can launch students on a journey of discovery. By exploring the patterns of one to four squares, students will have a basis for making predictions and finding all of the five-square patterns or pentominoes. No formula has yet been derived for mathematically determining how many polyominoes are possible for a given number of connected squares.

# of squares	# of patterns	Patterns
1 (monomino)	1	
2 (domino)	1	
3 (tromino)	2	
4 (tetromino)	5	
5 (pentomino)	12	
6 (hexomino)	35	

Rules for joining squares
1. Squares must touch along one entire edge.

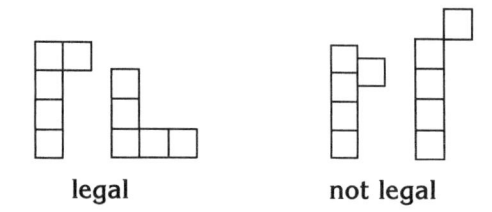

    legal   not legal

2. If a pattern can fit on top of another using a flip or turn, it is considered to be the same. For example, the patterns below are the same.

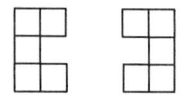

HARDHATTING IN A GEO-WORLD   © 2004 AIMS Education Foundation

Management

1. Make copies of the square grid on colored construction paper, one per student plus a few extra. Students should cut five individual squares from the construction paper or be provided with five square tiles.
2. Envelopes may be used to store pentomino pieces for later puzzle work or for doing the related activity, *Net-Sense*.
3. The overhead projector is a helpful tool for demonstrating the rules for joining squares as well as showing results. Have five squares available to manipulate on the projector.

(The following is offered for those students ready for more independent work.)

> *Open-ended:* Ask students, "In how many ways can one to five squares be joined together?" Give them the rules and five squares. Have them devise their own means of representing and reporting the data.

Procedure

1. Give students five square tiles or distribute the square grid and have students cut out five squares.
2. Ask the first *Key Question*, "In how many ways can squares be joined?" Demonstrate the rules for joining squares.
3. Once students have the first activity page, have them find and record all the ways to join one square.
4. Instruct students to move two of their squares into all of the possible positions and record. Repeat for three squares and four squares.
5. Ask students the second *Key Question*, "How many different patterns can be made with five squares?" Direct students to record their predictions.
6. Have students explore the five-square patterns and record by coloring in the 5x3 grids on the activity page. The emphasis is on discovery; do not give any clues as to how many pentominoes can be made. Toward the end of the discovery period, you may ask questions such as "Who can find the 13th pentomino?" so exploration will continue until the students are absolutely convinced all the pentominoes have been found. Encourage perseverance.
7. Direct students to use the square grid paper to cut out the pentomino pieces they have identified. Encourage them to conserve space. However some will need a second sheet to finish.
8. Discuss their findings. Have students hold up like pieces to make sure their set is complete. It is helpful to give each shape a letter name such as

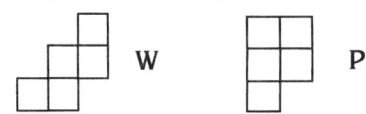

   W   P

9. Invite students to store pentominoes in envelopes for further investigations or extensions.

Connecting Learning

1. How many ways are there to make <u>one-square</u> patterns? [1] (...two-square? [1] ...three-square? [2] ...four-square? [5] ...five-square? [12])
2. How did you go about searching for all the five-square (pentomino) pieces? [moved the squares until I found something new, trial and error]
3. How could you conduct the pentomino search in an orderly way? [explore all the possibilities with five squares in a row, then four squares in a row, three squares in a row, etc.]
4. What are you wondering now?

Extensions

1. The area of every pentomino is the same—five square units. How do the perimeters compare? The edge of each square equals one unit.

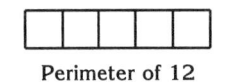

   Perimeter of 12

2. Challenge students to use all 12 pentomino pieces to make a rectangle, *not* a simple task. A 6 x 10 unit rectangle has over 1000 solutions. They can also try 5 x 12, 4 x 15, and 3 x 20 rectangles.

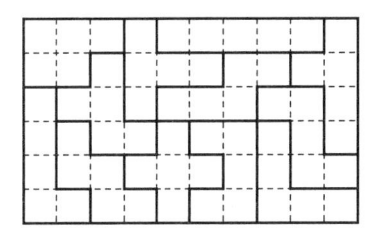

   One 6 x 10 solution

3. Explore the three-dimensional possibilities of pentominoes by doing the activity *Net-Sense*.
4. Have students find all of the hexominoes (six squares). Collect and display them.

Curriculum Correlation

Art
1. Suggest students trace the same pentomino piece four times to make a pattern pleasing to the eye. Does it have one or more lines of symmetry? Check with mirrors. Does it rotate around a center point?
2. Have students choose one pentomino piece and decide what it could become, perhaps by adding legs, leaves, wheels, etc. Trace around it and incorporate the shape into a picture. Write a story to go with the drawing.

Technology
   Have students write *Logo* programs for the pentomino shapes.

* Reprinted with permission from *Principles and Standards for School Mathematics,* 2000 by the National Council of Teachers of Mathematics. All rights reserved.

Edge to Edge

Key Questions

1. In how many ways can squares be joined together?
2. How many different patterns can be made with five squares?

Learning Goal

Students will:

- find as many ways as possible to combine one, two, three, four, and five squares, edge to edge.

Edge to Edge

HARDHATTING IN A GEO-WORLD   135   © 2004 AIMS Education Foundation

Edge to Edge

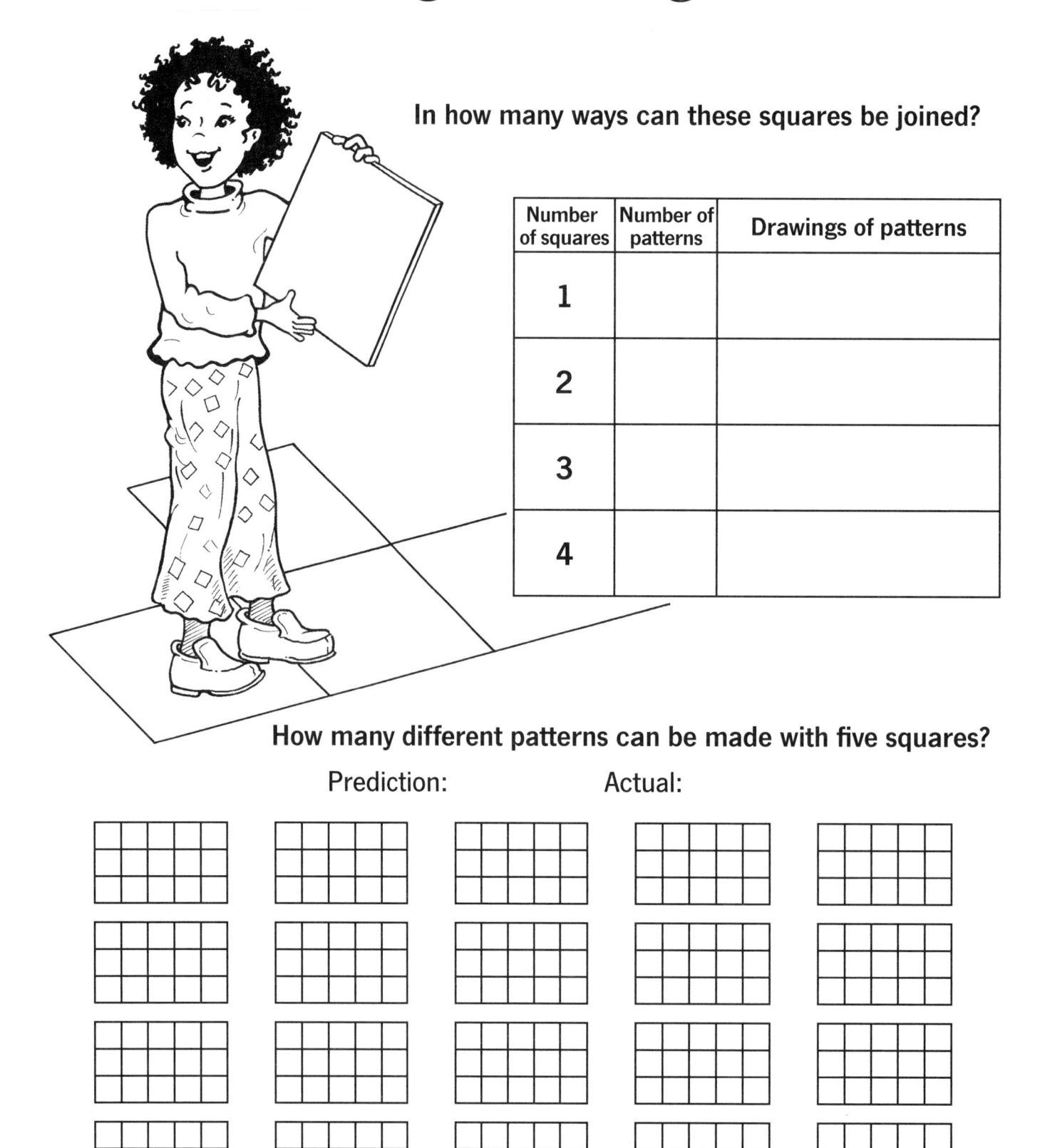

In how many ways can these squares be joined?

Number of squares	Number of patterns	Drawings of patterns
1		
2		
3		
4		

How many different patterns can be made with five squares?

Prediction: Actual:

HARDHATTING IN A GEO-WORLD   © 2004 AIMS Education Foundation

Edge to Edge

HARDHATTING IN A GEO-WORLD

Edge to Edge

1. How many ways are there to make <u>one-square</u> patterns? (…two-square? …three-square? …four-square? …five-square?)

2. How did you go about searching for all the five-square (pentomino) pieces?

3. How could you conduct the pentomino search in an orderly way?

4. What are you wondering now?

Net-Sense

Topic
Nets for boxes

Key Questions
1. Which pentomino nets can be folded into open boxes?
2. Which nets can be cut from an open milk carton?

Learning Goals
Students will:
- test which pentomino nets (2-D) fold into open boxes (3-D), and
- cut milk cartons (3-D) down into all the possible pentomino nets (2-D).

Guiding Document
NCTM Standards 2000*
- *Identify and build a three-dimensional object from two-dimensional representations of that object*
- *Identify and build a two-dimensional representation of a three-dimensional object*
- *Create and describe mental images of objects, patterns, and paths*

Math
Geometry and spatial sense
  2-D to 3-D
  3-D to 2-D
  spatial vizualization

Integrated Processes
Observing
Predicting
Collecting and recording data
Comparing and contrasting

Materials
Part One
  one set of pentominoes for each student
  scissors
  crayons

Part Two
  2 half-pint milk cartons for each student
  scissors
  crayons
  chart-sized paper for each group
  glue

Background Information
We want to give children many opportunities to build spatial knowledge—to explore the characteristics of two- and three-dimensional shapes, to see shapes in relationship to each other, and to examine the effects that changes have on shapes. The visual and tactile senses are employed as we construct, draw, compare, and transform things. Spatial visualization, the mental construction and manipulation of objects, is a component of spatial knowledge. This ability is being exercised and strengthened as students predict which pentominoes can be folded into open boxes and as they plan how to cut their milk carton down into pentomino nets.

A net is a pattern for building a shape or object. A one-dimensional net can be used to construct a two-dimensional shape. For example, connected lines form a square. A two-dimensional net creates a three-dimensional object. Certain pentominoes are two-dimensional nets for three-dimensional open boxes. A net creates an object one dimension higher than the net itself.

1-D net→2-D square   2-D net→3-D open box

Students will discover that eight pentominoes can be folded into open boxes. Conversely, the milk cartons can be cut down into the same eight pentomino nets. We want students to connect the folding up results of *Part One* with the cutting down results of *Part Two* and realize that they are just approaching the problem from two different points of view. If this link is made, they will know if they have found all the ways to cut down a milk carton or which ones they still need to find.

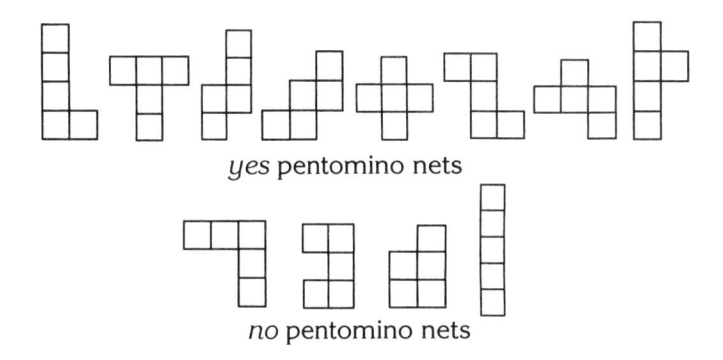

yes pentomino nets

no pentomino nets

HARDHATTING IN A GEO-WORLD

Management
1. Have students cut pentomino sets from the sheet provided.
2. You may wish to do *Part One* and *Part Two* on different days. Groups of four are desirable for *Part Two*.
3. Have students save their lunch milk cartons for a couple of days. The cartons should be rinsed and left to dry. Collect extra milk cartons, just in case they are needed.

Procedure
Part One
1. Have students gather their sets of pentominoes. Make sure everyone has all 12 pieces by holding up one pentomino at a time and having each student hold up the matching one.
2. Ask the first *Key Question* and distribute *Part One*. Explain that a net, in this case, is the pattern for making a three-dimensional figure.
3. Instruct students to **mentally** fold each pentomino pictured in the giraffe's neck into a box. To help with their visual thinking, suggest they put an *x* in the square they think would make the base. If they predict *yes*, have them color that pentomino piece on the paper.
4. Direct students to fold their paper pentominoes along the lines and draw the successful nets in the *yes* area and the unsuccessful nets in the *no* area.
5. Discuss the results and have students save their paper pentominoes for *Part Two*.

Part Two
1. Ask the second *Key Question* and distribute *Part Two* along with the milk cartons.
2. Have students cut the cartons so the sides are nearly square. Challenge them to trace a path that follows the edges or folds and will make the carton lie flat.
3. Instruct students to cut along their tracing until they have a flat pattern. They may have to add or subtract cuts from the path they traced.
4. Have students take their second milk carton and try to cut a different net.
5. Direct the groups to assemble their nets and compare. How many different nets did they make? Do they think there are others?
6. Give each group a large piece of paper and have they make a display of the *yes* paper pentominoes, the *no* paper pentominoes, and their milk carton nets.
7. Encourage each group to study their display, then answer the two questions on the activity page.

Connecting Learning
1. Compare the number of pentominoes that form open boxes with the total number of pentominoes. (Have students express their answer as a fraction, 8/12, and/or make a hand-drawn circle graph.)
2. Explain why you answered *no* to some of the nets.
3. How many cuts did you need to make to flatten the milk carton? [4] Did it always take the same number of cuts? [Yes.] Explain.
4. How many different milk carton nets were found? (There should be eight.) Did we find them all? How do you know? (See *Discussion 5.*)
5. How are the milk carton nets related to the folded paper pentominoes? (They should match because they are just the reverse process of each other. This is one way to find if any nets are missing among the cartons.)

Extensions
1. Explore lines of symmetry. Which pentominoes can be folded in half so that one side fits on the other? Mark the folds that work.
2. Design a net that will fold into a cube. Test it.

* Reprinted with permission from *Principles and Standards for School Mathematics*, 2000 by the National Council of Teachers of Mathematics. All rights reserved.

Net-Sense

Key Questions

1. Which pentomino nets can be folded into open boxes?
2. Which nets can be cut from an open milk carton?

Learning Goals

Students will:

- test which pentomino nets (2-D) fold into open boxes (3-D), and
- cut milk cartons (3-D) down into all the possible pentomino nets (2-D).

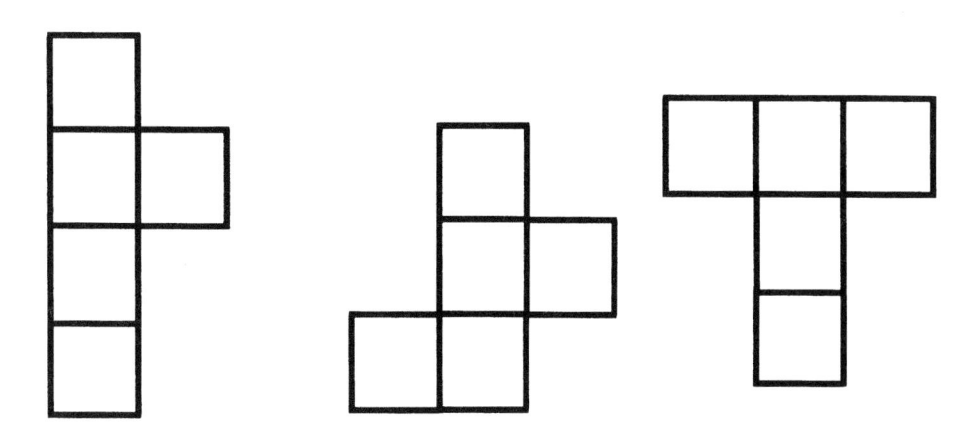

Net-Sense

Cut these pentomino nets.

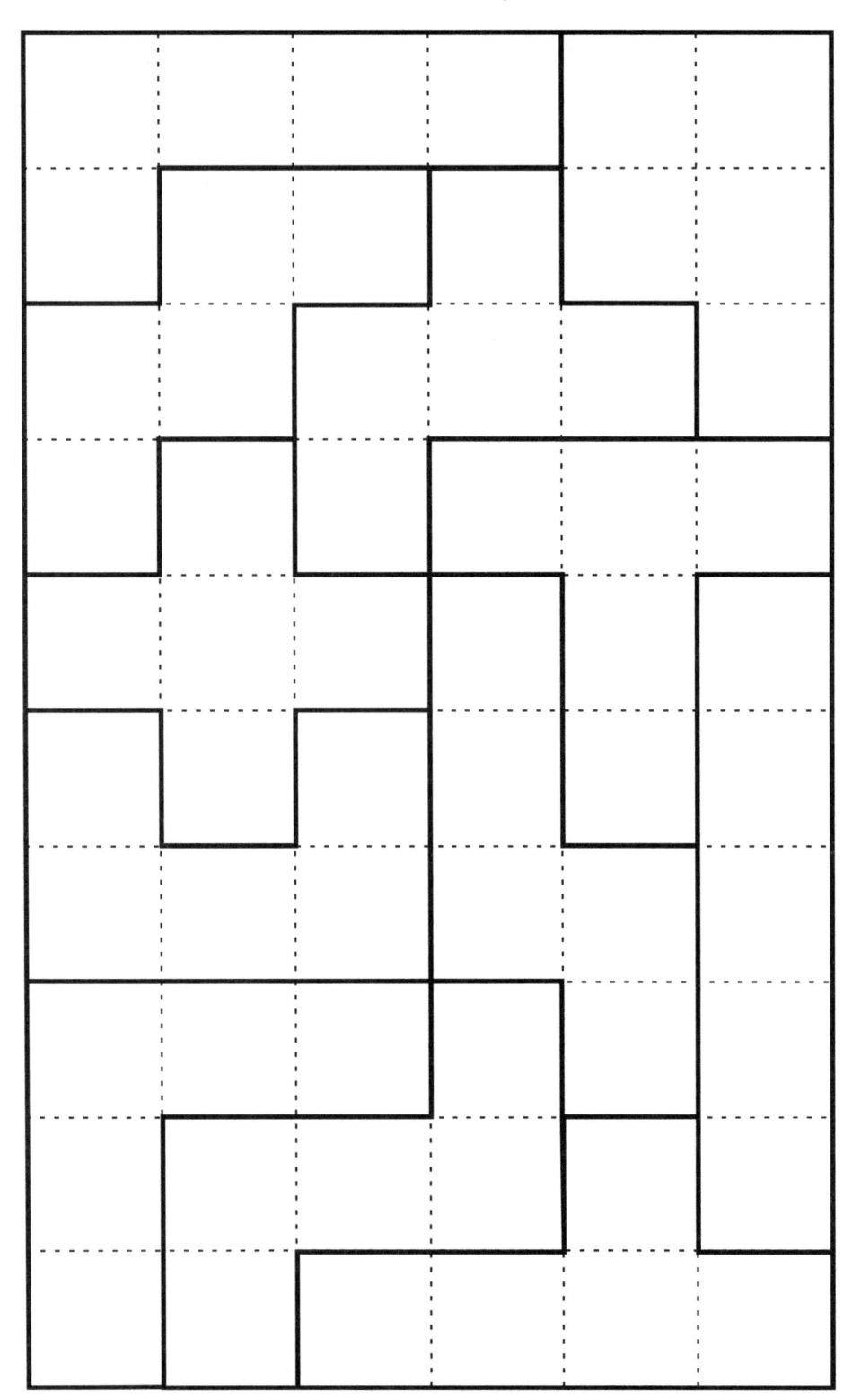

HARDHATTING IN A GEO-WORLD

Net-Sense
Part One

Predict by coloring.

Which pentomino nets can be folded into open boxes?

Yes

No

HARDHATTING IN A GEO-WORLD © 2004 AIMS Education Foundation

Net-Sense
Part Two

Which nets can be cut from an open milk carton?

With a crayon, trace the edges you want to cut on your milk carton. Cut and draw the net you made.

Use another milk carton to make a different net. Draw it.

Working as a team...

How many different nets did your group make?

Make a display of your three collections:
- pentominoes that form boxes
- pentominoes that do not form boxes
- milk carton nets

What do you observe?

What new questions do you have?

HARDHATTING IN A GEO-WORLD

Net-Sense

1. Compare the number of pentominoes that form open boxes with the total number of pentominoes.

2. Explain why you answered *no* to some of the nets.

3. How many cuts did you need to make to flatten the milk carton? Did it always take the same number of cuts? Explain.

4. How many different milk carton nets were found? Did we find them all? How do you know?

5. How are the milk carton nets related to the folded paper pentominoes?

Wreck-Tangles

Topic
Perimeter/area of rectangles

Key Question
How do the areas of rectangles with equal perimeters compare?

Learning Goals
Students will:
* compare the areas of rectangles with equal perimeters, and
* possibly learn a short cut (formula) for finding the area of a rectangle.

Guiding Documents
Project 2061 Benchmarks
* Length can be thought of as unit lengths joined together, area as a collection of unit squares, and volume as a set of unit cubes.
* Use numerical data in describing and comparing objects and events.

NCTM Standards 2000*
* Use geometric models to solve problems in other areas of mathematics, such as number and measurement
* Explore what happens to measurements of a two-dimensional shape such as its perimeter and area when the shape is changed in some way.

Math
Measurement
 length
 area
Geometry and spatial sense
Order

Integrated Processes
Observing
Collecting and recording data
Comparing and contrasting
Generalizing

Materials
For each group:
 about a 32 cm length of string
 4 pushpins
 1 cardboard box (see *Management 2*)
 transparent tape

Background Information
This activity is designed to enrich students' concepts of perimeter and area. To give students a tactile experience with perimeter, they start by maneuvering string into different rectangular positions on a centimeter grid. As students move to drawing rectangles on paper, various approaches may be observed. Some will draw rectangles with a given perimeter by trial and error, erasing lines as they go. Others will realize that one length plus one width equals half of the perimeter they want. Building upon their experience with the string perimeter, some might draw rectangles in an organized way. For instance, for a perimeter of 12, they might start with a 1 x 5 rectangle, then a 2 x 4 rectangle, etc. Let students make the connections naturally; some will not be ready to move out of the trial-and-error mode.

Having examined the data, students may think they have the definitive answer such as *rectangles with equal perimeters do not have equal areas* or *length times width equals area*. Their generalizations require further testing. Will the same results be true for perimeters of 24? Can examples be found that do not confirm their generalizations?

One of the big ideas of mathematics is maximum and minimum. The length and width that are the same or closest to being the same give the maximum area for a given perimeter. In contrast, the minimum area is formed where the length and width differ by the greatest amount. Using whole numbers and a perimeter of 12, the largest area is formed by a 3 x 3 rectangle and the smallest area is formed by a 1 x 5 rectangle.

Management
1. The teacher may wish to pre-tie the string into 30-centimeter loops. Colored string or crochet thread is a nice contrast to the white paper.
2. Beforehand, have each group of three bring a cardboard box whose bottom is a little larger than an 8 1/2" x 11" piece of paper.
3. *Part One* uses centimeters as the unit of measure. The perimeters in *Part Two*, are measured in units, one unit being the edge of a square. Each square, then, represents one square unit of area.
4. *Part Two* is designed for repeated use. A good initial perimeter length to explore is 12. Then try other even numbers such as 18, 24, or whatever you wish.
5. To find the area, students should count the squares inside each rectangle. (The area formula may be one of the insights gained from this activity; it should not be imposed on students as they are in the process of discovery.)

HARDHATTING IN A GEO-WORLD

Procedure

Part One
1. Distribute the pre-tied 30-cm string loops to each group, along with *Part One* and four push pins.
2. Explain that the string could represent a fence around a property, a perimeter. Have students take turns using two fingers of each hand to make different kinds of rectangles with the string.
3. Ask, "Did the length of your perimeter change when you made the different rectangles?" [No.] "I wonder if the amount of space (area) inside the rectangles changed. Let's find out!" 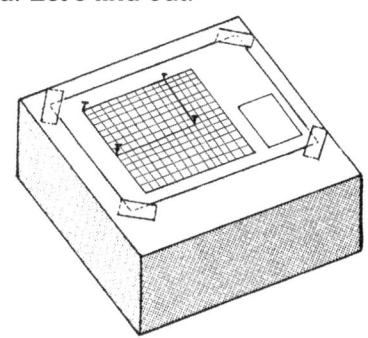
4. Have them turn their cardboard box upside-down and tape the paper to the bottom of it. Direct them to insert a pushpin where it says, "Start here." This pin should not be moved during the activity.
5. Instruct students to put the string loop around the base of the pushpin and put the other pushpins at centimeter intersections so that the thinnest rectangle possible is formed. Have them record length, width, and area.
6. Have students continue to make other rectangles by moving the three pushpins to different positions.
7. Discuss the results. Is there an organized way to find all the possible rectangles? Did anyone do it that way? What patterns do you see in the table?

Part Two
1. Distribute *Part Two* and decide what perimeter length to explore.
2. Have students draw as many different rectangles with the given perimeter as possible. Students should label the length, width, and area for each rectangle. 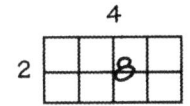
3. Challenge students to find a way to order their rectangles and record them in the table in this order. (They might order by increasing length, increasing width, or increasing area.) If they need more room, have them use the back of the paper.
4. Hold a concluding discussion.

Connecting Learning

1. How did you go about finding different rectangles with perimeters of ___? (Have students reflect on the process they used: trial and error, finding one length plus one width that equal half of the perimeter, drawing rectangles in an orderly way, etc.)
2. How many different rectangles did you make?
3. Do any of your rectangles match? (Matches will occur if a 1 x 5 and a 5 x 1 are considered to be different rectangles.) Explain how they match. [Their areas are the same. If you cut one rectangle out, it would fit on top of the other rectangle.]
4. Do you consider a 2 x 4 rectangle the same or different from a 4 x 2 rectangle? Explain. (Accept any reasonable answers. As a class, decide which definition will be used when examining the results and drawing conclusions.)
5. How did you order your rectangles in the table? In what other ways could they be ordered? [by increasing length, increasing width, or increasing area]
6. What other patterns do you see in your table? [length x width = area; the closer the rectangle is to a square, the larger its area, etc.]
7. If you had the materials for ____ (name the perimeter number used) meters of fencing, what rectangle would give you the largest enclosed area? [the one closest to a square shape] What rectangle would give you the smallest enclosed area? [The one where the length and width differ by the greatest amount.] Which would you rather have for a play area?
8. What conclusions can you make based on your data? (Examples: Rectangles with equal perimeters do not have equal areas. Length times width equals area. When the length and width are the same, the area will be the largest for a given perimeter. The length and width that differ by the greatest amount will have the smallest area for a given perimeter.) In what ways could you test your conclusions?

Extensions

1. Repeat this activity with a different perimeter. This allows continued testing of generalizations and reinforces their feel for the area formula.
2. Investigate whether a rectangle can be made from a odd-numbered perimeter.

* Reprinted with permission from *Principles and Standards for School Mathematics*, 2000 by the National Council of Teachers of Mathematics. All rights reserved.

Wreck-Tangles

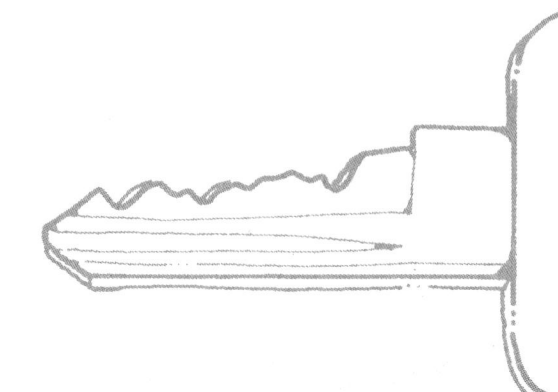

Key Question

How do the areas of rectangles with equal perimeters compare?

Learning Goals

Students will:

- compare the areas of rectangles with equal perimeters, and
- possibly learn a short cut (formula) for finding the area of a rectangle.

Wreck-Tangles
-Part One

How do the areas of rectangles with equal perimeters compare?

Make a string loop with a perimeter of 30 cm. Tape this paper to a box and use push pins to make different rectangles.

Length	Width	Area

Start here.

HARDHATTING IN A GEO-WORLD

Wreck-Tangles
Part Two

How do the areas of rectangles with equal perimeters compare?

Order the data and record.

Length	Width	Area

Draw rectangles with perimeters of _____.

What did you discover?

HARDHATTING IN A GEO-WORLD   150   © 2004 AIMS Education Foundation

Wreck-Tangles

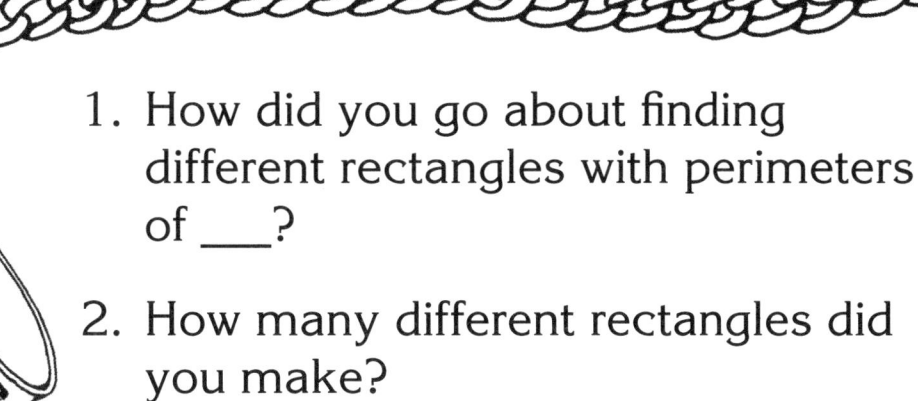

CONNECTING LEARNING

1. How did you go about finding different rectangles with perimeters of ___?

2. How many different rectangles did you make?

3. Do any of your rectangles match? Explain.

4. Do you consider a 2 x 4 rectangle the same or different from a 4 x 2 rectangle? Explain.

5. How did you order your rectangles in the table? How else could they be ordered?

6. What other table patterns do you see?

7. If you had the materials for ___ (perimeter number used) meters of fencing, what rectangle would give you the largest enclosed area? What rectangle would give you the smallest enclosed area? Which would you rather have for a play area?

8. What conclusions can you make based on your data? In what ways could you test your conclusions?

Paper Pinchers

Topic
Exploring area through origami

Key Questions
1. Challenge: Find all the possible ways to fold a square in half.
2. How does the area of a square change as it is folded?

Learning Goals
Students will:
- fold squares in various ways, and
- determine how the areas are related to each other.

Guiding Documents

Project 2061 Benchmarks
- Mathematics is the study of many kinds of patterns, including numbers and shapes and operations on them. Sometimes patterns are studied because they help to explain how the world works or how to solve practical problems, sometimes because they are interesting in themselves.
- Length can be thought of as unit lengths joined together, area as a collection of unit squares, and volume as a set of unit cubes.
- Measurements are always likely to give slightly different numbers, even if what is being measured stays the same.

NRC Standards
- Employ simple equipment and tools to gather data and extend the senses.
- Communicate investigations and explanations.

NCTM Standards 2000*
- Use geometric models to solve problems in other areas of mathematics, such as number and measurement
- Investigate, describe, and reason about the results of subdividing, combining, and transforming shapes
- Explore what happens to measurements of a two-dimensional shape such as its perimeter and area when the shape is changed in some way

Math
Measurement
  length
  area
Area formula
Geometry and spatial sense
Patterns

Integrated Processes
Observing
Predicting
Collecting and recording data
Comparing and contrasting
Drawing conclusions

Materials

For each student:
  several squares of thin paper (see *Management 1*)
  metric ruler

For the teacher:
  optional: wax paper squares

Background Information

Geometry is rich with patterns pulled from the real world yet, too often, it occupies a subservient position to other strands of mathematics. Some of its power and activity-based potential can be realized through the use of origami, an art to which students are naturally attracted.

A square piece of paper is the traditional base for origami projects. Here we investigate the effect folds have on the area of a square—first with half folds, then with folds that make increasingly smaller squares. The tactile experiences with geometric patterns, linked to the number patterns generated by measurement, provide students with a helpful foundation for building concepts.

Origami literally means paper folding (*ori*—to fold, *gami*—paper). In its purest form, no cutting or pasting is allowed. The idea of folding paper originated in ancient China, and became a truly creative art in the hands of the Japanese. It was first used to make elaborate paper ornaments to attach to ceremonial gifts *(noshi)* but has now become more of a recreational activity. Today origami enthusiasts are scattered throughout the world. New folds and creative ways to use origami continue to be discovered.

Japanese children are often taught origami when they are five or six years old. A child who becomes proficient is challenged to use increasingly smaller pieces of paper such as candy wrappers.

Management
1. Cut paper squares from gift wrapping paper, copy paper, or graph paper. Consider graph paper with an even number of squares for *Part One*, a 20-cm square for *Part Two*, and any size or kind (including origami paper) for *Part Three*.
2. Practice making the origami figures ahead of time.
3. A valley fold goes inward. A mountain fold forms a peak.

valley fold   mountain fold

HARDHATTING IN A GEO-WORLD   152   © 2004 AIMS Education Foundation

4. Most students will benefit from a step-by-step visual demonstration of the origami folds. Waxed paper squares, folded on an overhead projector, effectively show each step.

Procedure

Part One: Explorations with one fold
1. Distribute the first activity page and square paper.
2. Present the *Challenge:* Find all the possible ways to fold a square in half. (To keep it open-ended, there are more recording squares than needed.) 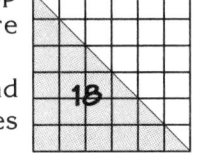
3. Have students shade, count, and record the number of grid squares on one side of the fold line.
4. Instruct students to write a statement comparing the areas of the shaded parts along with an explanation.
5. Challenge students to find another way to prove that the shaded areas are the same.

Part Two: Explorations with sets of folds
1. Ask the *Key Question,* "How does the area of a square change as it is folded?"
2. Distribute the second activity page and new square paper (or students can reuse their squares).
3. Direct students to measure the length and width of the open square to the nearest millimeter and find the area by an appropriate method.
4. Have students make the first set of folds, then measure and record. Repeat for the second and third sets of folds.
5. Instruct students to study the table and record any patterns that are found.
6. Using the patterns, have students predict the length, width, and area of a fourth fold.

Part Three: Explorations with origami
1. Give each student one of the origami pages and a square paper.
2. Have students fold the origami figure by following the page directions and your step-by-step demonstration.
3. Repeat for other origami figures, as desired.

Connecting Learning

Part One: Explorations with one fold
1. How many ways did you find to fold a square in half? [4] Show them.

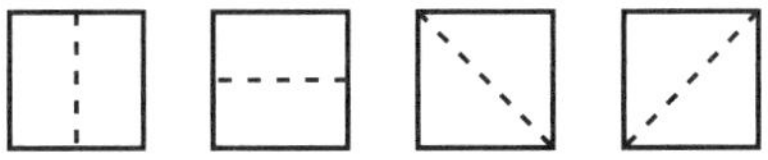

2. What do these fold lines tell us about the symmetry of a square? [The fold lines are lines of symmetry; a square has four.]

3. How can you prove the shaded areas are the same? [count the number of shaded squares; cut the triangle so that it can be superimposed on the rectangle; measure the rectangle and triangle and use formulas to find their areas] 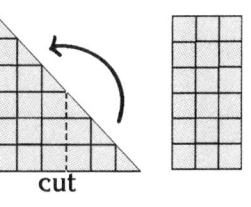
4. How does the perimeter change when the square is folded in half? Does it change in the same way that the area does?

Part Two: Explorations with sets of folds
1. How do the measurements within the class compare? What might cause differences? [how carefully the square was folded, how carefully students measured]
2. What patterns did you find in the table? [In every other set of folds, the length and width are cut in half. The area is reduced by half with each set of folds.] *The pattern may not be perfect, but should be close enough to detect.*
3. What should the measurements of the fourth fold be? [length and width should be half of the second set of folds, area should be half of the third set of folds]

Part Three: Explorations with origami
1. What kinds of geometric shapes can you find in your origami figures?
2. Which figure was the easiest to make? ...most difficult?
3. Who folded a figure with the smallest piece of paper? What size was your paper?

Extensions
1. Find the total area of the servant.
2. Calculate how many square pieces of paper had to be cut for the class to do this activity.
3. Discuss the kinds of triangles made by the folds.
4. Use a protractor to measure the angles of the various folded shapes.

Curriculum Correlation
Art
   Try other origami or paper-folding projects.

Literature
   Coerr, Eleanor. *Sadako and the Thousand Paper Cranes.* Puffin books. New York. 1999. After reading this book, learn how to fold the crane, a classic origami figure.

Home Link
   Have students teach family members to make one of the origami figures.

* Reprinted with permission from *Principles and Standards for School Mathematics,* 2000 by the National Council of Teachers of Mathematics. All rights reserved.

Paper Pinchers

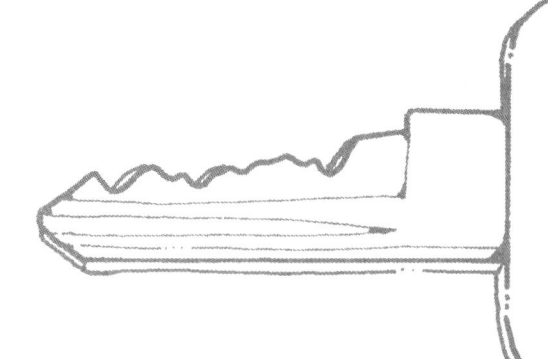

Key Questions

1. Challenge: Find all the possible ways to fold a square in half.
2. How does the area of a square change as it is folded?

Learning Goals

Students will:

- fold squares in various ways, and
- determine how the areas are related to each other.

Paper Pinchers
Explorations with One Fold

Challenge: Find all the possible ways to fold a square in half.

For each solution, draw the fold line and shade one side of the fold.
To find the area, count and record the shaded squares.

What did you notice? How would you explain that?

In what other ways (besides counting) can you show how the areas of the folded shapes are related?
Draw or describe a way that works.

HARDHATTING IN A GEO-WORLD   155   © 2004 AIMS Education Foundation

Paper Pinchers
Explorations with Sets of Folds

How does the area of a square change as it is folded?

Measure and record the dimensions of the open square. To fold, bring the corners to the center. Measure and record the dimensions of the folded square. Turn the square over and repeat. Make three sets of folds.

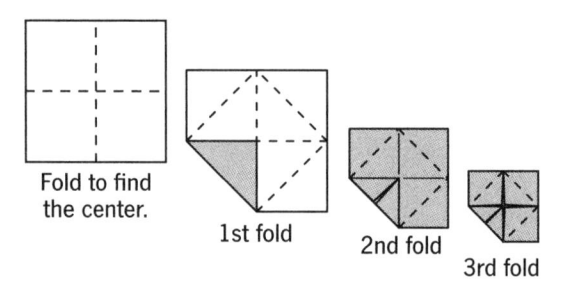

	Length (cm)	Width (cm)	Area (cm²)
Open square			
1st fold			
2nd fold			
3rd fold			

Study the table. Then predict the measurements of the square after the fourth fold.

4th fold			

What patterns do you see?

HARDHATTING IN A GEO-WORLD © 2004 AIMS Education Foundation

Paper Pinchers
Explorations with Origami

Servant

If using origami paper, start with the white side facing you.
(Drawings are not to scale.)

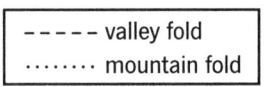

```
----- valley fold
······· mountain fold
```

1. To find the center of the square, fold in half twice. Unfold.

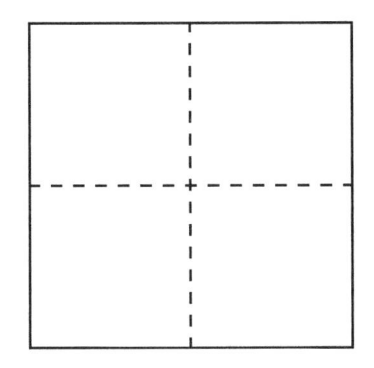

2. Fold each corner to the center.

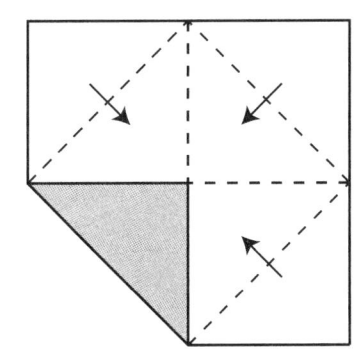

3. Turn the square over. Again, fold each corner to the center.

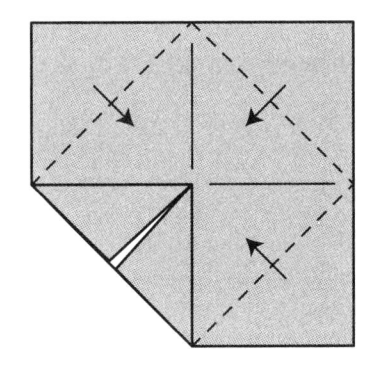

4. Turn the square over. Once more, fold each corner to the center.

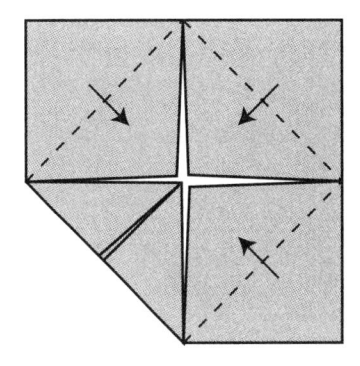

5. Turn the square over. Open and flatten three of the small squares by gently pulling up and out from the center slits.

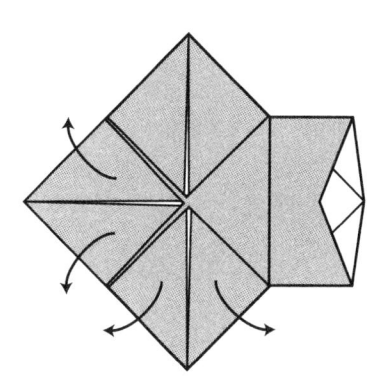

6. Draw a face on the remaining square.

HARDHATTING IN A GEO-WORLD © 2004 AIMS Education Foundation

Paper Pinchers
Explorations with Origami

Frame
If using origami paper, start with the colored side facing you.
(Drawings are not to scale.)

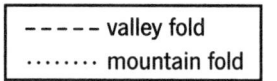

```
----- valley fold
......... mountain fold
```

1. To find the center of the square, fold in half twice. Unfold.

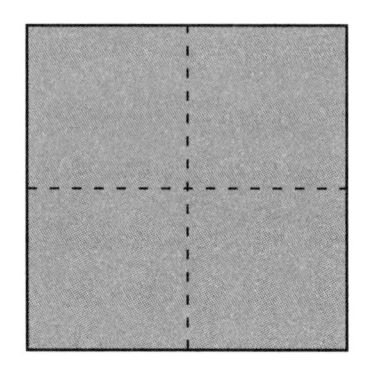

2. Fold each corner to the center.

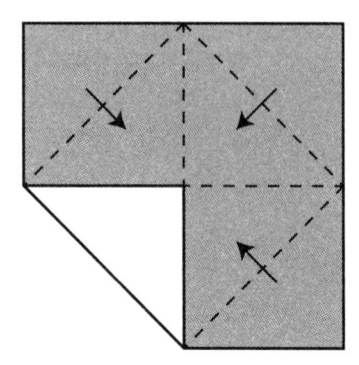

3. Turn the square over. Again, fold each corner to the center.

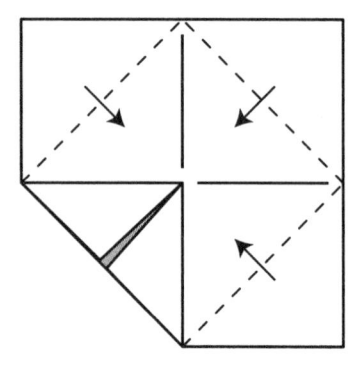

4. Turn the square over. Fold each square flap back to form a triangle.

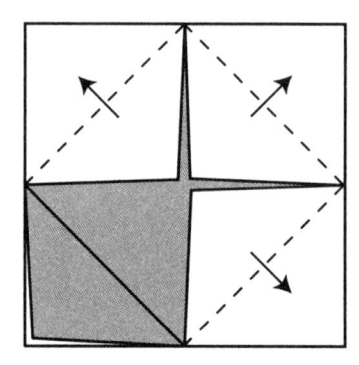

5. Slide a poem or picture under the corner triangles.

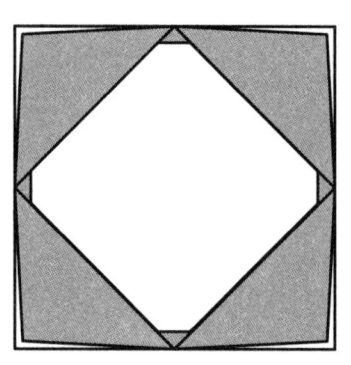

6. Stand the frame on one of the triangles in the back. For more support, unfold the right and left back triangles so they are at right angles to the frame.

HARDHATTING IN A GEO-WORLD © 2004 AIMS Education Foundation

Paper Pinchers
Explorations with Origami

Salt Cellar

If using origami paper, start with the white side facing you.
(Drawings are not to scale.)

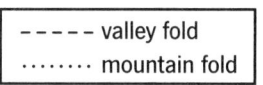

- - - - - valley fold
······· mountain fold

1. To find the center of the square, fold in half twice. Unfold.

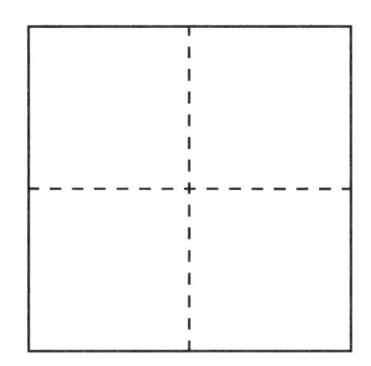

2. Fold each corner to the center.

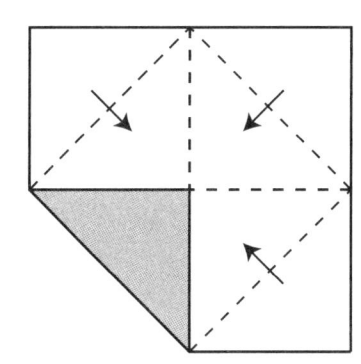

3. Turn the square over. Again, fold each corner to the center.

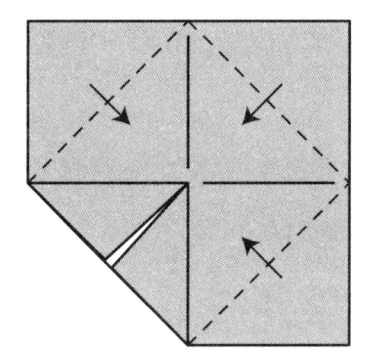

4. Turn the square over. Fold in half each way. Open back to the square shown below.

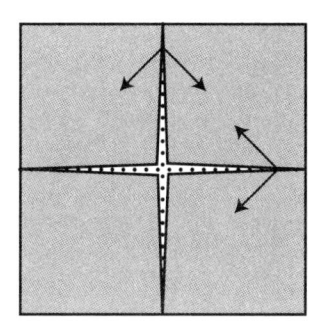

5. Put one finger under the center point. With your other hand, slide a finger into each of the four corners and push down to open.

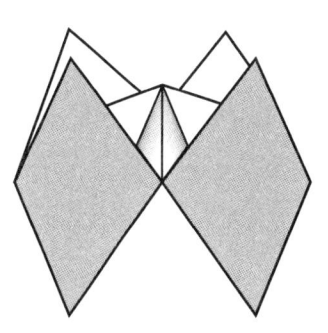

6. Fill your "dish" with raisins, sunflower seeds, or other goodies.

HARDHATTING IN A GEO-WORLD © 2004 AIMS Education Foundation

Paper Pinchers

Part One: Explorations with one fold
1. How many ways did you find to fold a square in half? Show them.
2. What do these fold lines tell us about the symmetry of a square?
3. How can you prove the shaded areas are the same?
4. How does the perimeter change when the square is folded in half? Does it change in the same way that the area does?

Part Two: Explorations with sets of folds
1. How do the measurements within the class compare? What might cause differences?
2. What patterns did you find in the table?
3. What should the measurements of the fourth fold be?

Part Three: Explorations with origami
1. What kinds of geometric shapes can you find in your origami figures?
2. Which figure was the easiest to make? ...most difficult?
3. Who folded a figure with the smallest piece of paper? What size was your paper?

Circle Sighs

Topic
Circles, radius and diameter

Key Question
What are the radius and diameter of each of your circles?

Learning Goal
Students will use paper clips to draw circles, determining their radii and diameters.

Guiding Documents
Project 2061 Benchmarks
- When people care about what is being counted or measured, it is important for them to say what the units are (three degrees Fahrenheit is different from three centimeters, three miles from three miles per hour).
- Tables and graphs can show how values of one quantity are related to values of another.
- Graphical display of numbers may make it possible to spot patterns that are not otherwise obvious, such as comparative size and trends.

NCTM Standards 2000
- Build and draw geometric objects
- Use geometric models to solve problems in other areas of mathematics, such as number and measurement
- Measure with multiple copies of units of the same size, such as paper clips laid end to end

Math
Geometry
 circles
Measurement

Integrated Processes
Observing
Comparing and contrasting
Collecting and recording data
Interpreting data
Drawing conclusions

Materials
Part One
 paper clips, jumbo and regular size
 large construction paper 12" x 18", 2 sheets per pair of students
 tape
 chart paper

Part Two
 paper clips, jumbo and regular size
 markers or colored tissue paper, white glue, and paint brushes (see *Management 1*)

Background Information
Before students begin to use compasses and protractors, they should be given time to explore and enjoy the drawing of circles. In this activity they will use paper clips to draw circles and paper clip chains to measure the diameter and radius of each circle. Through the collection of data, students should reach the conclusion that the diameter is twice the length of the radius.

Management
1. This activity is divided into two parts. In *Part One* the students will learn how to use paper clips to draw circles. They will collect and record data to determine the radius and diameter of each circle. In *Part Two* students will use paper clips to draw circles in interesting designs. They can then color their work or cut circles from colored tissue paper and adhere them to their drawings by applying a wash of white glue and water with a paint brush (one part white glue to two parts water).
2. Students should work in pairs with both taking turns drawing the circles.
3. Each pair will need 20 paper clips. They will use 10 chained together to make a measuring device with paper clip units (pc). The other ten will be used for drawing circles. For making comparisons, have some students use jumbo-size clips while others use regular-size clips.

Procedure
Part One
1. Ask the students how they could use a paper clip to draw a circle. As they are describing their methods, try to follow their directions on the overhead projector. If students do not arrive at the solution that is illustrated (or a better one), demonstrate how to place a pencil in each loop of the paper clip and trace a circle as they spin one end around the stationary pencil in the other loop.

HARDHATTING IN A GEO-WORLD © 2004 AIMS Education Foundation

2. Inform them that they will be drawing circles on a piece of paper, along with measuring and recording the radii and diameters of the circles they have drawn. If appropriate, weave the story of the ranch hands riding from the rim of Circle Ranch to the resting place in the middle of the ranch for the radius measure and all the way across the circle, passing by the resting place, for the diameter measure.
3. Show students a chain of paper clips. Explain that they will use a similar device to measure their circles. Ask what units they think they will record for their measures. [paper clip units, or pc]
4. Distribute a sheet of paper (scratch paper will do) to each pair of students. Allow them time to practice drawing the circles.
5. Distribute the construction paper, two sheets per student pair. Direct students to tape a long edge of one sheet to a long edge of the other.
6. Have the students devise a strategy for finding the center of the paper and draw a dot there. Allow time for them to share their strategies and assist any who are having difficulties. (Strategies: measure; fold the paper in half horizontally and vertically and use the intersection; draw diagonals with a straight edge and use the intersection)
7. Distribute 20 paper clips to each pair of students. Have them chain 10 together as a measuring device and practice reading the length of items around their area. (For example, they might report their pencil is three paper clip units long.)
8. After measuring practice, direct students to use the center dot to draw all of their circles. Have them draw a circle using only one paper clip, then measure and record the distance from the center to the edge of the circle (radius). Next, direct them to measure and record the distance from one edge to the other edge, making sure their measuring chain goes through the center of the circle (diameter).
9. Continue to draw, measure, and record with two and three paper clips. Then ask students to predict, without measuring, how many paper clips they can use before the circle gets too big for the paper. Have them record their reasoning.
10. Allow time for the students to continue until they can no longer add paper clips for drawing circles and remain on the paper. (Those using the smaller paper clips will have more data to collect.)
11. When all circles are drawn, have the students compare their data, explaining why some groups drew more than others. Have them record any patterns they observed in their data. [The radius is always twice as long as the diameter.]
12. Challenge the students to determine the size paper needed for a circle with a radius of 15 jumbo paper clips or 20 regular clips. Then invite them to draw the circle on that size of chart paper.

Part Two
1. Introduce some of the circle designs that are illustrated. Talk about the strategies for making them. Inform the students that the dots represent the middle of the circles they will draw. Allow students time to draw some.
2. Encourage the students to replicate a given design or create one of their own. Invite them to color their creations (or use the tissue paper/white glue wash).
3. Display their work.

Connecting Learning
1. What is the radius of a circle? [It is a straight line from the center point of a circle to the circumference.]
2. What is the diameter of a circle? [It is a straight line that passes through the center point of a circle from one edge to the other edge.]
3. How are the radius and diameter of a circle related? [The diameter is twice the length of the radius.]
4. Will this pattern be different if you use the other loop on the paper clip? Explain. [No, it still takes two radii (on opposite sides of the circle center) to make the diameter.]
5. How many circles were you able to draw on the construction paper? What does the difference in the numbers tell you about the paper clips? [We used two different sizes of paper clips.]
6. How are your circles alike? [They are all round. They share the same center—they're concentric.] How are they different? [The radius and diameter differ in each one.]
7. What patterns did you notice in your data?
8. What patterns do you get when circles of the same size are drawn so that their centers are placed on the circumference of another circle? What do you notice? What is the measure between centers? What about circles of different sizes?
9. What things did you discover when you were drawing your circle designs? Which designs do you want to explore further? Why?

Extensions
1. Have students symmetrically color their circle designs.
2. Have students research the five rings of the Olympic symbol and replicate them.
3. Investigate drawing circles using loops of string instead of paper clips.

* Reprinted with permission from *Principles and Standards for School Mathematics*, 2000 by the National Council of Teachers of Mathematics. All rights reserved.

Circle Sighs

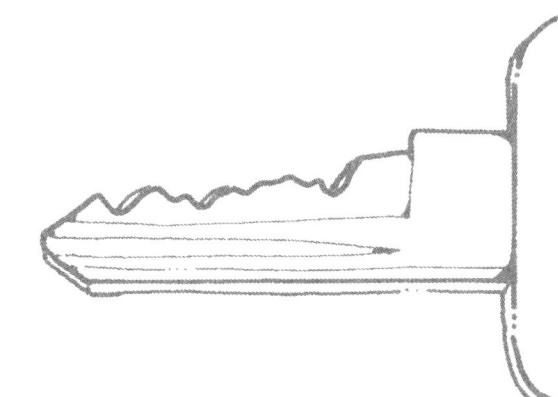

Key Question

What are the radius and diameter of each of your circles?

Learning Goal

Students will:

- use paper clips to draw circles, determining their radii and diameters.

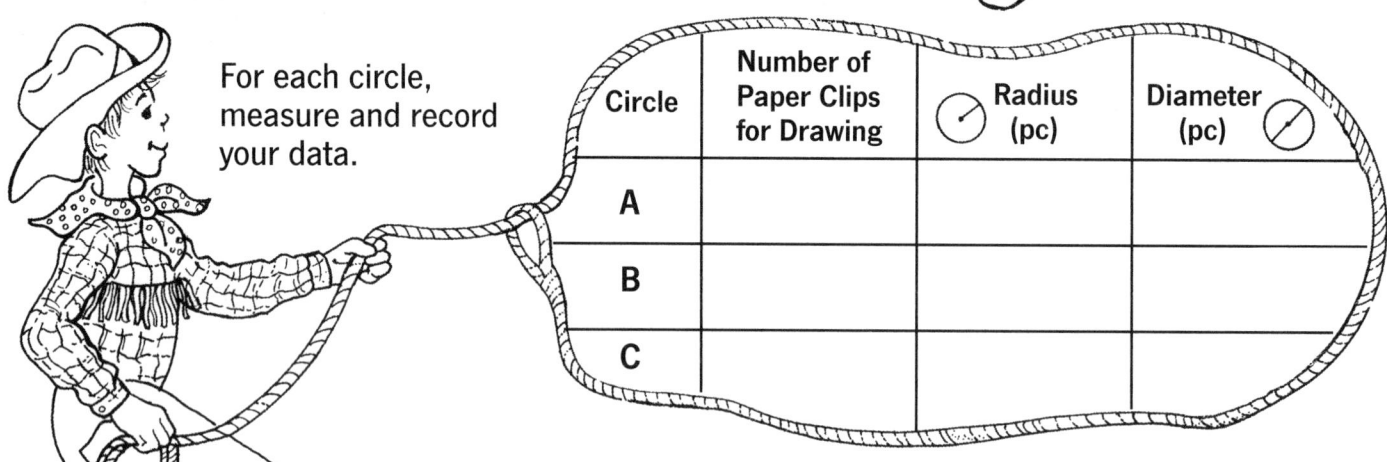

Circle Sighs

For each circle, measure and record your data.

Circle	Number of Paper Clips for Drawing	Radius (pc)	Diameter (pc)
A			
B			
C			

Without measuring, predict how many paper clips you'll be able to use for drawing circles before you go off the paper.

How did you decide?

Finish collecting your data.

Circle	Number of Paper Clips for Drawing	Radius (pc)	Diameter (pc)
D	4		
E	5		
F	6		
G	7		
H	8		
I	9		
J	10		

What patterns do you notice in your data?

HARDHATTING IN A GEO-WORLD © 2004 AIMS Education Foundation

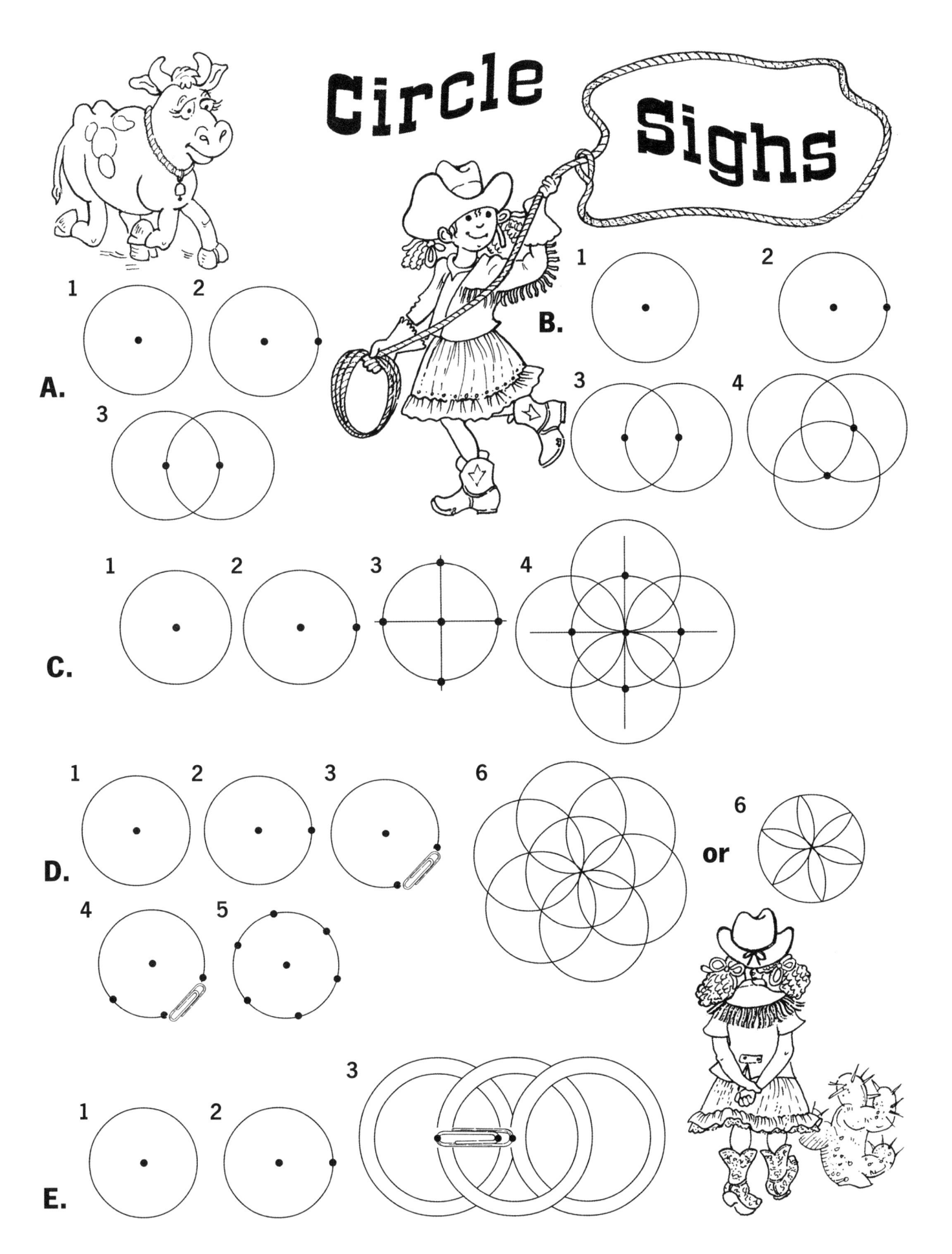

Circle Sighs

A.
1, 2, 3

B.
1, 2, 3, 4

C.
1, 2, 3, 4

D.
1, 2, 3, 4, 5, 6 or 6

E.
1, 2, 3

HARDHATTING IN A GEO-WORLD 165 © 2004 AIMS Education Foundation

Circle Sighs

1. What is the radius of a circle?

2. What is the diameter of a circle?

3. How are the radius and diameter of a circle related?

4. Will this pattern be different if you use the other loop on the paper clip? Explain.

5. How many circles were you able to draw on the construction paper? What does the difference in the numbers tell you about the paper clips?

6. How are your circles alike? How are they different?

7. What patterns did you notice in your data?

8. What patterns do you get when circles of the same size are drawn so that their centers are placed on the circumference of another circle? What do you notice? What is the measure between centers? What about circles of different sizes?

9. What things did you discover when you were drawing your circle designs? Which designs do you want to explore further? Why?

Playground Geometry

Topic
Geometry/measurement

Key Question
How do the measurements of geometric shapes on our playground compare?

Learning Goal
Students will show what they know about geometric words and relationships by measuring squares, rectangles, and circles on their playground.

Guiding Documents
Project 2061 Benchmarks
- Shapes such as circles, squares, and triangles can be used to describe many things that can be seen.
- Measurements are always likely to give slightly different numbers, even if what is being measured stays the same.

NRC Standard
- Employ simple equipment and tools to gather data and extend the senses.

NCTM Standards 2000*
- Recognize geometric ideas and relationships and apply them to other disciplines and to problems that arise in the classroom or in everyday life.
- Select and apply appropriate standard units and tools to measure length, area, volume, weight, time, temperature, and the size of angles

Math
Geometry and spatial sense
Measurement
　　linear
Rounding
Whole number operations
　　addition: perimeter
　　multiplication: area

Integrated Processes
Observing
Grouping
Collecting and recording data
Comparing and contrasting

Materials
Meter sticks
Trundle wheels or meter tapes
Calculators

Background Information
　　We can come to appreciate the many geometric shapes in our environment by deliberately drawing our attention to them. And that is what students are asked to do in this activity as they focus on the school playground environment.
　　The tactile experiences involved in measuring length, width, diameter, etc. enrich students' concepts of these terms. It also allows them to look for relationships among the parts of shapes. For example, a radius is half the length of a diameter.
　　It is appropriate for students at this level to identify and measure circumference. Discussing the relationship of circumference to diameter or using circumference to find area should be deferred until they are developmentally ready.

Management
1. Students should have previous experiences with length, width, perimeter, area, radius, diameter, and circumference before doing this activity. It could be used as a culmination or for assessment.
2. The whole activity can be done at one time or squares and rectangles can be done one day, circles on another.
3. You may wish to have each group record data on one activity sheet or for each individual in the group to keep a record of the data.
4. Choose whether you want everyone to measure the same objects or permit each small group to choose their own objects to measure.
5. If everyone is measuring the same objects, have each group start in a different place and rotate from object to object.
6. Finding the radius or diameter may not be easy unless it is marked or students know where the center of the circle is. If it is not marked and is on a flat surface, mark it for them with chalk. Otherwise they will need to approximate it.

HARDHATTING IN A GEO-WORLD　　　　167　　　　© 2004 AIMS Education Foundation

Procedure

(The following is the suggested procedure to use if the whole class will be measuring the same objects. See *Management* for another option and modify accordingly.)

1. Challenge students to find and measure squares, rectangles, and circles on their playground. Take the class outside for some initial scouting. Encourage them to offer their suggestions, then agree on the objects to be used.
2. Distribute the activity sheet and measuring tools to each group. Have students record the objects to be measured. Draw out their understanding of terms. Example: "Describe a perimeter." [the distance around a closed shape; *peri* (around) *meter* (measure)]
3. Discuss the procedure for gathering data (rotation or every group on their own). Ask questions such as:
   - From where shall we start measuring thick lines—the outside edge, the inside edge, or the middle of the line?
   - How shall we round our measurements, to the nearest centimeter or the nearest meter? (Large areas like soccer fields should be rounded to the nearest meter, smaller areas to the nearest centimeter.)
4. Take the class outside. Have the groups measure and record their data.
5. Return to the room so students can do perimeter and area calculations.
6. For circles, have students create their own table or other way of recording data. One possibility is:

Object	Radius	Diameter	Circumference

7. Have groups go outside and obtain the circle data.
8. Hold a concluding discussion.

Connecting Learning

1. What problems did you have while you were gathering data?
2. How do our measurements of _____ (name object) compare? Why might they be different? (Some students measured more precisely than others or from the outer or inner edge of a thick line, etc.)
3. How is perimeter related to length and width? [length + width + length + width or 2 lengths + 2 widths]
4. What is another word for the free throw line? [diameter]
5. How does the radius compare to the diameter? Use data to support your answer. (The radius is half the diameter. For example,...)
6. How are perimeter and circumference alike? [*Peri* and *circum* both mean "around." They are both distances around a closed shape. Circumference is just a word chosen to represent a special kind of perimeter.] How are they different? [Perimeters have straight lines and circumferences are curved.]
7. While you were outside, what other geometric shapes did you notice?

* Reprinted with permission from *Principles and Standards for School Mathematics,* 2000 by the National Council of Teachers of Mathematics. All rights reserved.

Playground Geometry

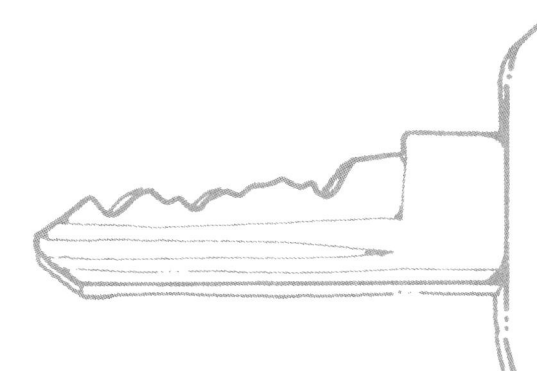

Key Question

How do the measurements of geometric shapes on our playground compare?

Learning Goal

Students will:

- show what they know about geometric words and relationships by measuring squares, rectangles, and circles on their playground.

Playground Geometry

Squares and Rectangles

Measure length and width. Calculate perimeter and area.

Object	Length	Width	Perimeter	Area

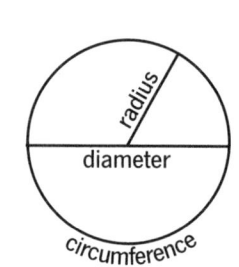

Circles

Measure and record the radius, diameter, and circumference of three circles.

Playground Geometry

1. What problems did you have while you were gathering data?

2. How do our measurements of _____ compare? Why might they be different?

3. How is perimeter related to length and width?

4. What is another word for the free throw line?

5. How does the radius compare to the diameter? Use data to support your answer.

6. How are perimeter and circumference alike? How are they different?

7. While you were outside, what other geometric shapes did you notice?

Once Around the Track

Topic
Geometry: closed networks

Key Question
When can a network be traced without lifting your pencil?

Learning Goal
Students will discover the rules that determine whether a network can be drawn without lifting a pencil or retracing a line.

Guiding Documents
Project 2061 Benchmarks
- *Mathematics is the study of many kinds of patterns, including numbers and shapes and operations on them. Sometimes patterns are studied because they help to explain how the world works or how to solve practical problems, sometimes because they are interesting in themselves.*
- *Results of scientific investigations are seldom exactly the same, but if the differences are large, it is important to try to figure out why. One reason for following directions carefully and for keeping records of one's work is to provide information on what might have caused the differences.*
- *Offer reasons for their findings and consider reasons suggested by others.*

NRC Standards
- *Employ simple equipment and tools to gather data and extend the senses.*
- *Communicate investigations and explanations.*

NCTM Standards 2000*
- *Recognize geometric ideas and relationships and apply them to other disciplines and to problems that arise in the classroom or in everyday life.*
- *Build new mathematical knowledge through problem solving*
- *Describe, extend, and make generalizations about geometric and numeric patterns*

Math
Geometry and spatial sense
Patterns

Integrated Processes
Observing
Collecting and recording data
Comparing and contrasting
Generalizing

Materials
None

Background Information
A Swiss mathematician by the name of Leonhard Euler (1707-1783) was fascinated by networks and discovered rules that determine whether or not networks can be traveled. A network consists of a set of points or vertices connected by edges which may be straight or curved. A network can be traveled if it can be traced without lifting a pencil or retracing an edge.

Network	# of Even Vertices	# of Odd Vertices	Can it be traveled?
A	0	2	yes
B	5	0	yes
C	1	4	no
D	3	2	yes
E	0	4	no
F	4	2	yes
G	6	0	yes
H	0	6	no
I			

Euler's discoveries about closed networks relevant to this activity are:
1. *A network can be traveled if it has two or fewer odd vertices.* An *odd vertex* has an odd number of edges or lines drawn from it. An *even vertex* has an even number of lines drawn from it.

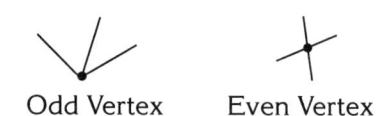

   Odd Vertex   Even Vertex

2. *The starting point sometimes matters. To travel a network with two odd vertices, you must start at one of the odd vertices. You will end at the other odd vertex.* (Until students find this truth, there is likely to be a difference in some of their results. Encourage students to retest the networks where their results don't agree. It can give them a tremendous sense of accomplishment to find the key. This is also why it is important to mark the starting point.)

The number of even vertices has no effect on which networks can be traveled.

HARDHATTING IN A GEO-WORLD © 2004 AIMS Education Foundation

Management
Students should work individually but might discuss and compare in small groups.

Procedure
1. Distribute the first activity sheet and ask, "Which of these networks can be traced without lifting your pencil or retracing a line?"
2. Have students trace each network and write *yes* or *no* on the line by its identifying letter. They should draw and test a network of their own for letter *I*.
3. Give students the second activity sheet and ask the *Key Question*, "When can a network be traced without lifting your pencil?"
4. Explain that a vertex is a point where two or more lines meet. Review the definitions of *odd* and *even* vertices as illustrated on the sheet.
5. Have students complete the table, either individually or in small groups.
6. Guide the reporting of results (counts of vertices and networks that can be traveled) so students can make comparisons.
7. Students should retest networks where results differed. They should find that the starting point sometimes determines whether a network can be traveled or not (see *Background Information*). Their thinking should be recorded under the question, "Why might someone get different results?"
8. Have students study the odd vertices and write a general rule to answer the *Key Question*.

Connecting Learning
1. What rules did you write? How did you test it to make sure the rule was a good one? (Students should try additional networks of their own to see if their rule holds true.)
2. What do you notice about the numbers in the odd vertices column? [They are all even numbers.] Do you think this will always be true? [For closed networks, yes. (But challenge students to prove otherwise.)]
3. How many different networks that can be traveled can we find? (see *Extension*)

Extension
Start a bulletin board showing networks that can be traveled. Have students draw them on colored construction paper and glue yarn along the lines to add more color. As the display grows, challenge students to find new networks that can be traveled.

* Reprinted with permission from *Principles and Standards for School Mathematics*, 2000 by the National Council of Teachers of Mathematics. All rights reserved.

Once Around the Track

Key Question

When can a network be traced without lifting your pencil?

Learning Goal

Students will:

- discover the rules that determine whether a network can be drawn without lifting a pencil or retracing a line.

Once Around the Track

When can a network be traced without lifting your pencil?

Mark your starting point (vertex). Trace lines only one time. Write **yes** if the network can be traveled, **no** if it cannot.

A _____

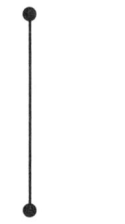

B _____

C _____

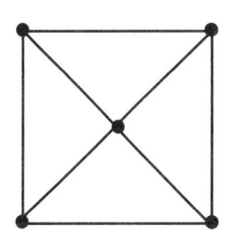

D _____

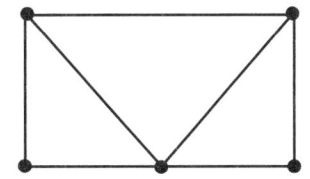

E _____

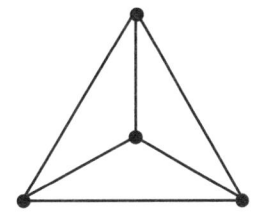

F _____

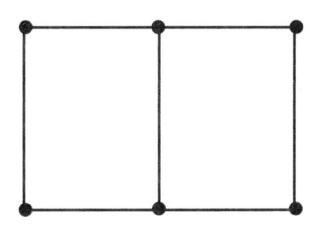

G _____

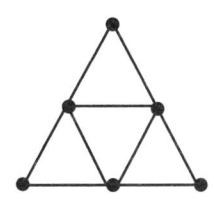

H _____

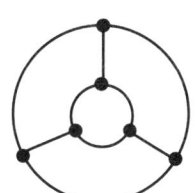

I _____

Draw your own.

HARDHATTING IN A GEO-WORLD   175   © 2004 AIMS Education Foundation

Once Around the Track

An **odd** vertex has an odd number of lines drawn from it.

An **even** vertex has an even number of lines drawn from it.

Network	# of even vertices	# of odd vertices	Can it be traveled?
A			
B			
C			
D			
E			
F			
G			
H			
I			

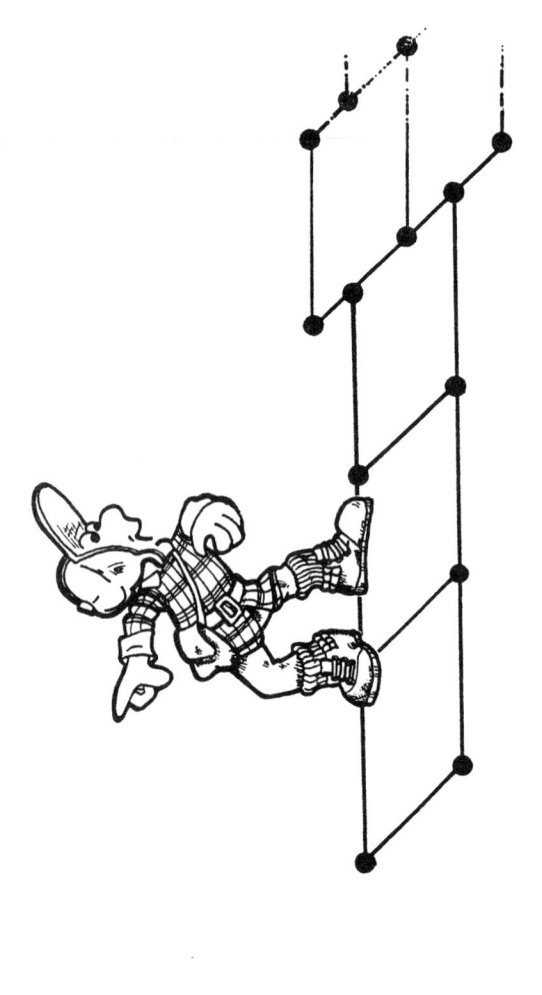

How do your results compare to others?

Why might someone get different results?

Look at the number of odd vertices in networks that can be traveled. Write a rule.

HARDHATTING IN A GEO-WORLD © 2004 AIMS Education Foundation

Once Around the Track

Connecting Learning

1. What rule(s) did you write? How did you test it to make sure the rule was a good one?

2. What do you notice about the numbers in the odd vertices column? Do you think this will always be true?

3. How many different networks that can be traveled can we find?

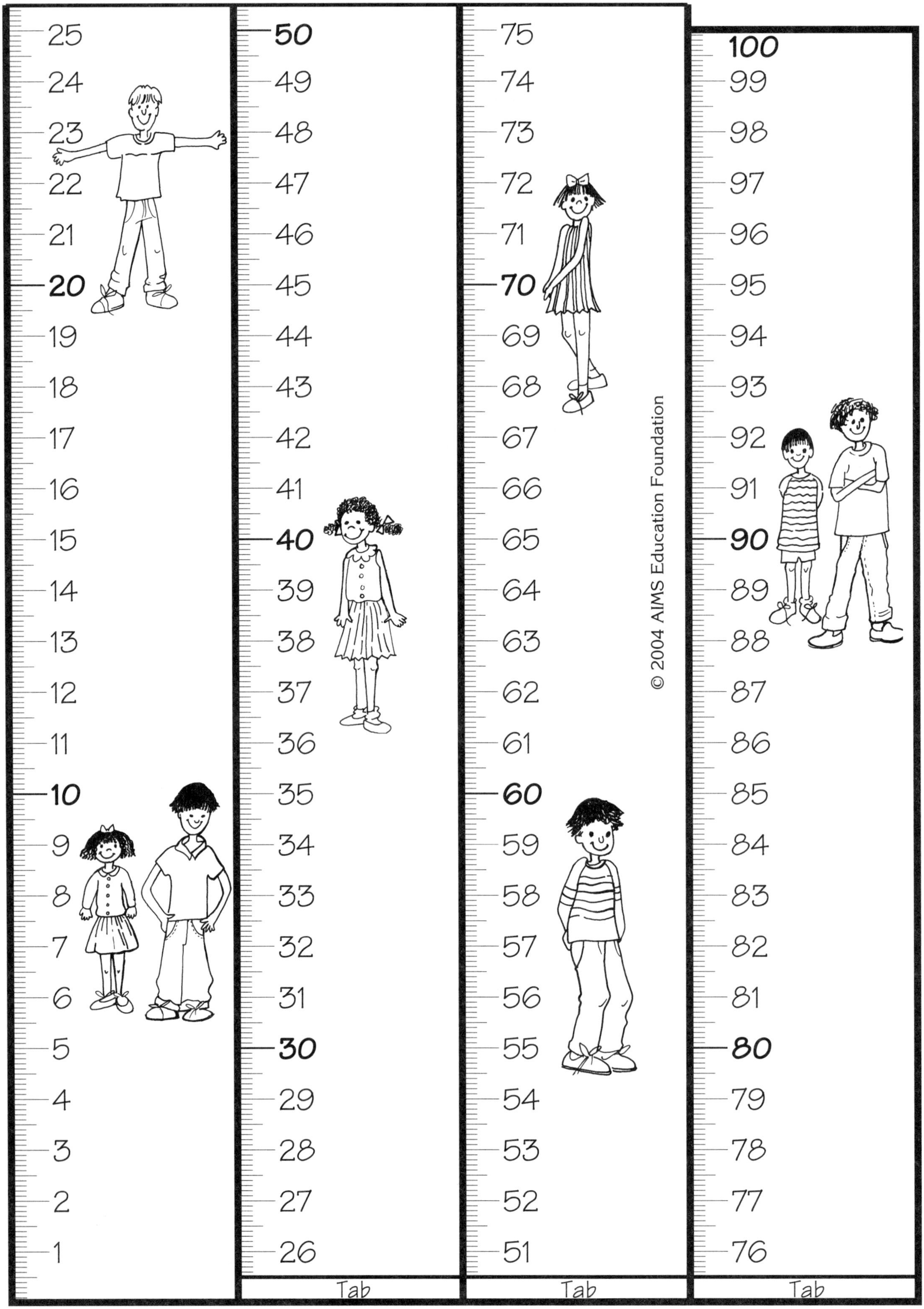
25
24
23
22
21
20
19
18
17
16
15
14
13
12
11
10
9
8
7
6
5
4
3
2
1
50
49
48
47
46
45
44
43
42
41
40
39
38
37
36
35
34
33
32
31
30
29
28
27
26
Tab
75
74
73
72
71
70
69
68
67
66
65
64
63
62
61
60
59
58
57
56
55
54
53
52
51
Tab
100
99
98
97
96
95
94
93
92
91
90
89
88
87
86
85
84
83
82
81
80
79
78
77
76
Tab

The AIMS Program

AIMS is the acronym for "**A**ctivities **I**ntegrating **M**athematics and **S**cience." Such integration enriches learning and makes it meaningful and holistic. AIMS began as a project of Fresno Pacific University to integrate the study of mathematics and science in grades K-9, but has since expanded to include language arts, social studies, and other disciplines.

AIMS is a continuing program of the non-profit AIMS Education Foundation. It had its inception in a National Science Foundation funded program whose purpose was to explore the effectiveness of integrating mathematics and science. The project directors in cooperation with 80 elementary classroom teachers devoted two years to a thorough field-testing of the results and implications of integration.

The approach met with such positive results that the decision was made to launch a program to create instructional materials incorporating this concept. Despite the fact that thoughtful educators have long recommended an integrative approach, very little appropriate material was available in 1981 when the project began. A series of writing projects have ensued, and today the AIMS Education Foundation is committed to continue the creation of new integrated activities on a permanent basis.

The AIMS program is funded through the sale of books, products, and staff development workshops and through proceeds from the Foundation's endowment. All net income from program and products flows into a trust fund administered by the AIMS Education Foundation. Use of these funds is restricted to support of research, development, and publication of new materials. Writers donate all their rights to the Foundation to support its on-going program. No royalties are paid to the writers.

The rationale for integration lies in the fact that science, mathematics, language arts, social studies, etc., are integrally interwoven in the real world from which it follows that they should be similarly treated in the classroom where we are preparing students to live in that world. Teachers who use the AIMS program give enthusiastic endorsement to the effectiveness of this approach.

Science encompasses the art of questioning, investigating, hypothesizing, discovering, and communicating. Mathematics is the language that provides clarity, objectivity, and understanding. The language arts provide us powerful tools of communication. Many of the major contemporary societal issues stem from advancements in science and must be studied in the context of the social sciences. Therefore, it is timely that all of us take seriously a more holistic mode of educating our students. This goal motivates all who are associated with the AIMS Program. We invite you to join us in this effort.

Meaningful integration of knowledge is a major recommendation coming from the nation's professional science and mathematics associations. The American Association for the Advancement of Science in *Science for All Americans* strongly recommends the integration of mathematics,
science, and technology. The National Council of Teachers of Mathematics places strong emphasis on applications of mathematics such as are found in science investigations. AIMS is fully aligned with these recommendations.

Extensive field testing of AIMS investigations confirms these beneficial results:
1. Mathematics becomes more meaningful, hence more useful, when it is applied to situations that interest students.
2. The extent to which science is studied and understood is increased, with a significant economy of time, when mathematics and science are integrated.
3. There is improved quality of learning and retention, supporting the thesis that learning which is meaningful and relevant is more effective.
4. Motivation and involvement are increased dramatically as students investigate real-world situations and participate actively in the process.

We invite you to become part of this classroom teacher movement by using an integrated approach to learning and sharing any suggestions you may have. The AIMS Program welcomes you!

AIMS Education Foundation Programs

Practical proven strategies to improve student achievement

When you host an AIMS workshop for elementary and middle school educators, you will know your teachers are receiving effective usable training they can apply in their classrooms immediately.

Designed for teachers—AIMS Workshops:
- Correlate to your state standards;
- Address key topic areas, including math content, science content, problem solving, and process skills;
- Teach you how to use AIMS' effective hands-on approach;
- Provide practice of activity-based teaching;
- Address classroom management issues, higher-order thinking skills, and materials;
- Give you AIMS resources; and
- Offer college (graduate-level) credits for many courses.

Aligned to district and administrator needs—AIMS workshops offer:
- Flexible scheduling and grade span options;
- Custom (one-, two-, or three-day) workshops to meet specific schedule, topic and grade-span needs;
- Pre-packaged one-day workshops on most major topics—only $3,900 for up to 30 participants (includes all materials and expenses);
- Prepackaged *week-long* workshops (four- or five-day formats) for in-depth math and science training—only $12,300 for up to 30 participants (includes all materials and expenses);
- Sustained staff development, by scheduling workshops throughout the school year and including follow-up and assessment;
- Eligibility for funding under the Eisenhower Act and No Child Left Behind; and
- Affordable professional development—save when you schedule consecutive-day workshops.

University Credit—Correspondence Courses

AIMS offers correspondence courses through a partnership with Fresno Pacific University.
- Convenient distance-learning courses—you study at your own pace and schedule. No computer or Internet access required!

The tuition for each three-semester unit graduate-level course is $264 plus a materials fee.

The AIMS Instructional Leadership Program

This is an AIMS staff-development program seeking to prepare facilitators for leadership roles in science/math education in their home districts or regions. Upon successful completion of the program, trained facilitators become members of the AIMS Instructional Leadership Network, qualified to conduct AIMS workshops, teach AIMS in-service courses for college credit, and serve as AIMS consultants. Intensive training is provided in mathematics, science, process and thinking skills, workshop management, and other relevant topics.

Introducing AIMS Science Core Curriculum

Developed in alignment with your state standards, AIMS' Science Core Curriculum gives students the opportunity to build content knowledge, thinking skills, and fundamental science processes.
- *Each* grade specific module has been developed to extend the AIMS approach to full-year science programs.
- *Each* standards-based module includes math, reading, hands-on investigations, and assessments.

Like all AIMS resources these core modules are able to serve students at all stages of readiness, making these a great value across the grades served in your school.

For current information regarding the programs described above, please complete the following:

Information Request

Please send current information on the items checked:

____ *Basic Information Packet* on AIMS materials   ____ Hosting information for AIMS workshops
____ *AIMS Instructional Leadership Program*   ____ AIMS Science Core Curriculum

Name _____   Phone _____

Address_____
            Street   City   State   Zip

AIMS Magazine

YOUR K-9 MATH AND SCIENCE CLASSROOM ACTIVITIES RESOURCE

The AIMS Magazine is your source for standards-based, hands-on math and science investigations. Each issue is filled with teacher-friendly, ready-to-use activities that engage students in meaningful learning.

- *Four issues each year (fall, winter, spring, and summer).*

Current issue is shipped with all past issues within that volume.

1821	Volume XXI	2006-2007	$19.95
1822	Volume XXII	2007-2008	$19.95

Two-Volume Combination

M20507	Volumes XX & XXI	2005-2007	$34.95
M20608	Volumes XXI & XXII	2006-2008	$34.95

Back Volumes Available
Complete volumes available for purchase:

1802	Volume II	1987-1988	$19.95
1804	Volume IV	1989-1990	$19.95
1805	Volume V	1990-1991	$19.95
1807	Volume VII	1992-1993	$19.95
1808	Volume VIII	1993-1994	$19.95
1809	Volume IX	1994-1995	$19.95
1810	Volume X	1995-1996	$19.95
1811	Volume XI	1996-1997	$19.95
1812	Volume XII	1997-1998	$19.95
1813	Volume XIII	1998-1999	$19.95
1814	Volume XIV	1999-2000	$19.95
1815	Volume XV	2000-2001	$19.95
1816	Volume XVI	2001-2002	$19.95
1817	Volume XVII	2002-2003	$19.95
1818	Volume XVIII	2003-2004	$19.95
1819	Volume XIX	2004-2005	$19.95
1820	Volume XX	2005-2006	$19.95

Volumes II to XIX include 10 issues.

Call 1.888.733.2467 or go to www.aimsedu.org

Subscribe to the AIMS Magazine

$19.95 a year!

AIMS Magazine is published four times a year.

Subscriptions ordered at any time will receive all the issues for that year.

AIMS Online—www.aimsedu.org

To see all that AIMS has to offer, check us out on the Internet at www.aimsedu.org. At our website you can search our activities database; preview and purchase individual AIMS activities; learn about core curriculum, college courses, and workshops; buy manipulatives and other classroom resources; and download free resources including articles, puzzles, and sample AIMS activities.

AIMS News
While visiting the AIMS website, sign up for AIMS News, our FREE e-mail newsletter. You'll get the latest information on what's new at AIMS including:

- New publications;
- New core curriculum modules; and
- New materials.

Sign up today!

AIMS Program Publications

Actions with Fractions, 4-9
Awesome Addition and Super Subtraction, 2-3
Bats Incredible! 2-4
Brick Layers II, 4-9
Chemistry Matters, 4-7
Counting on Coins, K-2
Cycles of Knowing and Growing, 1-3
Crazy about Cotton, 3-7
Critters, 2-5
Electrical Connections, 4-9
Exploring Environments, K-6
Fabulous Fractions, 3-6
Fall into Math and Science, K-1
Field Detectives, 3-6
Finding Your Bearings, 4-9
Floaters and Sinkers, 5-9
From Head to Toe, 5-9
Fun with Foods, 5-9
Glide into Winter with Math and Science, K-1
Gravity Rules! 5-12
Hardhatting in a Geo-World, 3-5
It's About Time, K-2
It Must Be A Bird, Pre-K-2
Jaw Breakers and Heart Thumpers, 3-5
Looking at Geometry, 6-9
Looking at Lines, 6-9
Machine Shop, 5-9
Magnificent Microworld Adventures, 5-9
Marvelous Multiplication and Dazzling Division, 4-5
Math + Science, A Solution, 5-9
Mostly Magnets, 2-8
Movie Math Mania, 6-9
Multiplication the Algebra Way, 6-8
Off the Wall Science, 3-9
Out of This World, 4-8
Paper Square Geometry:
    The Mathematics of Origami, 5-12
Puzzle Play, 4-8
Pieces and Patterns, 5-9
Popping With Power, 3-5
Positive vs. Negative, 6-9
Primarily Bears, K-6
Primarily Earth, K-3
Primarily Physics, K-3
Primarily Plants, K-3
Problem Solving: Just for the Fun of It! 4-9
Problem Solving: Just for the Fun of It! Book Two, 4-9
Proportional Reasoning, 6-9
Ray's Reflections, 4-8
Sensational Springtime, K-2
Sense-Able Science, K-1
Soap Films and Bubbles, 4-9
Solve It! K-1: Problem-Solving Strategies, K-1
Solve It! 2nd: Problem-Solving Strategies, 2
Solve It! 3rd: Problem-Solving Strategies, 3
Solve It! 4th: Problem-Solving Strategies, 4
Solve It! 5th: Problem-Solving Strategies, 5
Spatial Visualization, 4-9
Spills and Ripples, 5-12
Spring into Math and Science, K-1
The Amazing Circle, 4-9
The Budding Botanist, 3-6
The Sky's the Limit, 5-9
Through the Eyes of the Explorers, 5-9
Under Construction, K-2
Water Precious Water, 2-6
Weather Sense: Temperature, Air Pressure, and Wind, 4-5
Weather Sense: Moisture, 4-5
Winter Wonders, K-2

Spanish Supplements*
Fall Into Math and Science, K-1
Glide Into Winter with Math and Science, K-1
Mostly Magnets, 2-8
Pieces and Patterns, 5-9
Primarily Bears, K-6
Primarily Physics, K-3
Sense-Able Science, K-1
Spring Into Math and Science, K-1

* Spanish supplements are only available as downloads from the AIMS website. The supplements contain only the student pages in Spanish; you will need the English version of the book for the teacher's text.

Spanish Edition
Constructores II: Ingeniería Creativa Con Construcciones LEGO® 4-9
    The entire book is written in Spanish. English pages not included.

Other Publications
Historical Connections in Mathematics, Vol. I, 5-9
Historical Connections in Mathematics, Vol. II, 5-9
Historical Connections in Mathematics, Vol. III, 5-9
Mathematicians are People, Too
Mathematicians are People, Too, Vol. II
What's Next, Volume 1, 4-12
What's Next, Volume 2, 4-12
What's Next, Volume 3, 4-12

For further information write to:
AIMS Education Foundation • P.O. Box 8120 • Fresno, California 93747-8120
www.aimsedu.org • 559.255.6396 (fax) • 888.733.2467 (toll free)